MEMOIRS
of the
American Mathematical Society

Number 450

Solution of a Non-domestic Tame Classification Problem from Integral Representation Theory of Finite Groups $(\Lambda = RC_3, \upsilon(3) = 4)$

Ernst Dieterich

July 1991 • Volume 92 • Number 450 (third of 4 numbers) • ISSN 0065-9266

American Mathematical Society
Providence, Rhode Island

1980 *Mathematics Subject Classification* (1985 *Revision*).
Primary 20C10; Secondary 16A64, 15A21.

Library of Congress Cataloging-in-Publication Data

Dieterich, Ernst, 1951–
Solution of a non-domestic tame classification problem from integral representation theory of finite groups ([Lambda] = RC_1, v(3) = 4)/Ernst Dieterich.
p. cm. – (Memoirs of the American Mathematical Society, ISSN 0065-9266; no. 450)
Includes bibliographical references.
ISBN 0-8218-2521-6
1. Representations of groups. 2. Finite groups. 3. Modules (Algebra) I. Title. II. Series.
QA3.A57 no. 450
[QA171]
510 s–dc20
[512′.2]

91-15023
CIP

Subscriptions and orders for publications of the American Mathematical Society should be addressed to American Mathematical Society, Box 1571, Annex Station, Providence, RI 02901-1571. *All orders must be accompanied by payment.* Other correspondence should be addressed to Box 6248, Providence, RI 02940-6248.

SUBSCRIPTION INFORMATION. The 1991 subscription begins with Number 438 and consists of six mailings, each containing one or more numbers. Subscription prices for 1991 are $270 list, $216 institutional member. A late charge of 10% of the subscription price will be imposed on orders received from nonmembers after January 1 of the subscription year. Subscribers outside the United States and India must pay a postage surcharge of $25; subscribers in India must pay a postage surcharge of $43. Expedited delivery to destinations in North America $30; elsewhere $82. Each number may be ordered separately; *please specify number* when ordering an individual number. For prices and titles of recently released numbers, see the New Publications sections of the NOTICES of the American Mathematical Society.

BACK NUMBER INFORMATION. For back issues see the AMS Catalogue of Publications.

MEMOIRS of the American Mathematical Society (ISSN 0065-9266) is published bimonthly (each volume consisting usually of more than one number) by the American Mathematical Society at 201 Charles Street, Providence, Rhode Island 02904-2213. Second Class postage paid at Providence, Rhode Island 02940-6248. Postmaster: Send address changes to Memoirs of the American Mathematical Society, American Mathematical Society, Box 6248, Providence, RI 02940-6248.

10 9 8 7 6 5 4 3 2 1 95 94 93 92 91

TABLE OF CONTENTS

Abstract

Suppose R is a complete discrete valuation ring, with exponential valuation v, and G is a finite p-group. In studying the representation type (finite, tame or wild) of the group ring RG, the last open problem was the case where $G = C_3$ and $v(3) = 4$. Here it is shown that this group ring is tame (non-domestic and of finite growth), by way of classifying all indecomposable representations of C_3 in R, when $v(3) = 4$.

Key words and phrases

Lattices over group rings, modules over finite-dimensional algebras, classification of indecomposable representations, tame representation type, tubular algebras, matrix reductions, reduction functors.

Received by the editors March 27, 1986.

INTRODUCTION

Let $\Lambda = RG$ be a group ring which is given by a complete discrete valuation ring R and a finite p-group G, such that Λ is an isolated singularity in the sense of Auslander [Au 84]. Here, this is equivalent to the condition that the prime number p is neither a unit nor zero in R. Consider the problem of classifying all indecomposable left Λ-lattices and, what is more, the problem of determining the Auslander-Reiten quiver of Λ.

The present state of knowledge in this respect can be summarized as follows. Denote by C_{p^n} the cyclic group of order p^n and by v the exponential valuation of R. Then Λ is of <u>finite representation type</u> if and only if one of the following cases (i)-(iv) occurs:

(i) $G = C_2$,

(ii) $G = C_3$ and $v(3) \leq 3$,

(iii) $G = C_p$ and $v(p) \leq 2$,

(iv) $G = C_{p^2}$ and $v(p) = 1$ [Dr/Ro 67], [Ja 67].

Moreover, Λ is of <u>tame representation type</u> in each of the following cases (v)-(vii):

(v) $G = C_2 \times C_2$ and $v(2) = 1$ [Na 67],

(vi) $G = C_8$ and $v(2) = 1$ [Ja 72],

(vii) $G = C_4$ and $v(2) = 2$ [Ko 75].

In each of the cases (i)-(v) the Auslander-Reiten quiver of Λ is known [Di 83a]. Furthermore, Λ is domestic tame in case (v), whereas it is non-domestic tame in cases (vi) and (vii).

(Recall that Λ is said to be of tame representation type if it is of infinite representation type, and if for each rank d there exists a finite collection $\{\mathfrak{M}_1^{(d)},\dots,\mathfrak{M}_{n_d}^{(d)}\}$ of one-parameter series of indecomposable Λ-lattices such that almost all indecomposable Λ-lattices of rank d belong to $\bigcup_{i=1}^{n_d} \mathfrak{M}_i^{(d)}$, up to isomorphism. In particular, Λ is said to be domestic tame if it is of infinite representation type and if there exists a finite collection $\{\mathfrak{M}_1,\dots,\mathfrak{M}_n\}$ of one-parameter series of indecomposable Λ-lattices such that for each rank d almost all indecomposable Λ-lattices of rank d belong to $\bigcup_{i=1}^{n} \mathfrak{M}_i$, up to isomorphism. On the other hand, Λ is said to be non-domestic tame if it is of tame representation type, but not domestic tame. Here, a one-parameter series of indecomposable Λ-lattices is understood to be a full subcategory $\mathfrak{M}$ of the category of all indecomposable Λ-lattices, together with a representation equivalence $F : \mathfrak{M} \longrightarrow R$, where R denotes the category of all indecomposable regular Kronecker modules over the residue class field k of R .)

Finally, Λ is of wild representation type if neither it belongs to (i)-(vii), nor to

(*) $G = C_3$ and $v(3) = 4$ [Di 83b] .

In view of these results the case (*) is the last open problem in the attempt to determine all group rings Λ of tame representation type within the given setting. It is this particular group ring for which the present article determines the Auslander-Reiten quiver, and thus solves the classification problem of its lattice category and proves that it is non-domestic tame.

There is another important aspect concerning the mathematical context of case (*). Consider the simple elliptic curve singularity of type $\tilde{\mathbb{E}}_8$, given by the polynomial $f(X,Y) = Y(Y-X^2)(Y-aX^2)$, $a \in \mathbb{C} \setminus \{0,1\}$, and let $\hat{\mathcal{O}} = \mathbb{C}[[X,Y]]/(f(X,Y))$ be its complete local ring. Consider the problem of classifying the isomorphism classes of indecomposable finitely generated torsionfree $\hat{\mathcal{O}}$-modules, respectively of giving a complete description for the Auslander-Reiten quiver of $\hat{\mathcal{O}}$. It turns out that this problem is completely analogous to the problem we are dealing with, in case the field of fractions of R is a splitting field for C_3 . It can be solved along the same line and, replacing I (see below) by $\mathbb{P}_1\mathbb{C}$, it has the same solution as is presented here. The significance of this observation lies in the fact that recently there has been increasing interest in the study of maximal Cohen--Macaulay modules over isolated hypersurface singularities. Whereas so far most results are concerned with hypersurface singularities of finite Cohen-Macaulay type, the above unimodular plane curve singularity seems to be the first example of an isolated hypersurface singularity of tame Cohen-Macaulay type for which a complete description if its Auslander-Reiten quiver is known.

With these remarks, which hopefully provide some motivation for the study of the classification problem of the lattice category over the particular group ring (*), we leave the general setting.

From now on let R be a complete discrete valuation ring with exponential valuation v , such that $v(3) = 4$, and let $\Lambda = RC_3$ be the group ring which is given by R and by the cyclic group of order 3. Denote by ${}_{\Lambda}L$ the category of all (left) Λ-lattices. In this article we solve two related problems.

(I) The problem of giving a full classification for the isomorphism classes of indecomposable Λ-lattices.

(II) The problem of giving a complete description for the Auslander--Reiten quiver of Λ .

There will be two solutions for each of these problems, depending on the further assumption whether the field of fractions K of R is a splitting field for C_3 or not. Nevertheless the two different solutions will be very closely connected.

We need a closer look at this distinction. Choose a fixed parameter π of R and let $k = R/\pi R$ be the residue class field of R . By hypothesis there is a uniquely determined unit $d \in R$ satisfying the equation $d\pi^4 + 3 = 0$. Let $\bar{d}$ be the corresponding residue class in k and consider the quadratic polynomial $\delta = X^2 - \bar{d}$ in $k[X]$. Then the following properties of R are equivalent.

(i) K is a splitting field for C_3 .

(ii) R contains a primitive third root of unity.

(iii) δ is reducible in $k[X]$.

Indeed, (i) and (ii) are equivalent by the Chinese Remainder Theorem. If R contains a primitive third root of unity α , then $\pi^{-2}(2\alpha + 1)$ is a zero in R of the quadratic polynomial $\tilde{\delta} = X^2 - d$ in $R[X]$, and therefore δ is reducible in $k[X]$. Conversely, if δ is reducible in $k[X]$ then $\tilde{\delta}$ is reducible in $R[X]$, by Hensel's Lemma, and therefore $\alpha = \frac{1}{2}(-1 + \pi^2\sqrt{d})$ is a primitive third root of unity in R . Thus (ii) and (iii) are equivalent.

Set $I = \{\lambda \in k[X] \mid \lambda$ is monic and irreducible$\} \cup \{\infty\}$. Consider the finite subset E of I which is given by $E = \{\infty, 0, \delta\}$ in case K is not a splitting field for C_3 , respectively by $E = \{\infty, 0, \zeta, -\zeta\}$ in case K is a splitting field for C_3 and $\delta = (X - \zeta)(X + \zeta)$. (Here and in the sequel we notationally identify a linear polynomial $X - \eta$

in I with its zero η .) For the solution of problems (I) and (II), the elements in E will play the rôle of "exceptional parameters". Let $\bar{I} = I \dot{\cup} \{\bar{\lambda} \mid \lambda \in E\}$ be the set which is obtained from I by adding another copy $\bar{\lambda}$ for each exceptional parameter λ . Then the solution to problem (I) can be formulated as follows.

THEOREM I. There is a bijection between the set of all isomorphism classes of indecomposable nonprojective Λ-lattices and the set $\mathbb{P}_1\mathbb{Q} \times \bar{I} \times \mathbb{N}$.

Moreover, it will become clear from the course of the proof that there is an effective algorithm for constructing the indecomposable Λ-lattice corresponding to any given triple in $\mathbb{P}_1\mathbb{Q} \times \bar{I} \times \mathbb{N}$ (see appendix), as well as for finding the invariants in $\mathbb{P}_1\mathbb{Q} \times \bar{I} \times \mathbb{N}$ corresponding to any given indecomposable nonprojective Λ-lattice. In this sense Theorem I does solve the classification problem for the category of Λ-lattices.

Before we can formulate the solution to problem (II) we have to collect a few notions in connection with the concept of Auslander-Reiten quiver. For later use it is convenient to do this at once for arbitrary Krull-Schmidt categories, including the category of Λ-lattices as a special case.

A category K is said to be R-additive, provided K has finite direct sums and all morphism sets $K(X,Y)$ are finitely generated R-modules such that the composition of morphisms $K(Y,Z) \times K(X,Y) \to K(X,Z)$ is R-bilinear, for all $X,Y,Z \in K$. An R-additive category is said to be a Krull-Schmidt category, provided all indecomposable objects have local endomorphism ring. Hence, if K is a Krull-Schmidt category then the decomposition of objects of K into indecomposable direct summands is unique up to isomorphism and order of occurence.

Let K be a Krull-Schmidt category. We denote by $\mathrm{ind}K$ the full subcategory consisting of all indecomposable objects of K. If $X \in K$ then $[X]$ denotes the isomorphism class of X in K. If $X \in \mathrm{ind}K$ then we set $f_X = K(X,X)/\mathrm{rad}K(X,X)$. By hypothesis, this is a skew field. With any pair of objects $(X,Y) \in \mathrm{ind}K \times \mathrm{ind}K$ we associate the following data: $\mathrm{rad}_K(X,Y)$ is the $(K(Y,Y),K(X,X))$-bimodule of all non-isomorphisms in $K(X,Y)$, $\mathrm{rad}_K^2(X,Y)$ is the subbimodule of $\mathrm{rad}_K(X,Y)$ generated by the set of morphisms $\bigcup_{Z \in \mathrm{ind}K} \mathrm{rad}_K(Z,Y)\mathrm{rad}_K(X,Z)$ and $\mathrm{irr}_K(X,Y)$ is the factorbimodule $\mathrm{rad}_K(X,Y)/\mathrm{rad}_K^2(X,Y)$. By definition, $\mathrm{irr}_K(X,Y)$ is an (f_Y,f_X)-bimodule, and we set a_{XY} to be its dimension as a left f_Y-vectorspace and a'_{XY} to be its dimension as a right f_X-vectorspace.

With any Krull-Schmidt category K we associate its Auslander--Reiten quiver $A(K)$. It is the valued oriented graph whose set of points is given by $\{[X] \mid X \in \mathrm{ind}K\}$ and whose set of arrows is defined by the rule that for any pair of points $([X],[Y])$ there exists an arrow $[X] \longrightarrow [Y]$ if and only if $\mathrm{irr}_K(X,Y) \neq 0$. Each arrow $[X] \longrightarrow [Y]$ is endowed with the valuation (a_{XY}, a'_{XY}). Following the usual convention we omit the valuation of an arrow in case it is equal to $(1,1)$.

If the Krull-Schmidt category K has Auslander-Reiten sequences then the Auslander-Reiten quiver $A(K)$ becomes a translation quiver with respect to the translation mapping τ, and the stable subquiver $A_s(K)$ of $A(K)$ is defined. The domain of τ is the set of all isomorphism classes of indecomposable objects in K which occur as end term of an Auslander-Reiten sequence of K. The description of this set depends on the type of the Krull-Schmidt category K. We refer to [Ri 84] for a detailed discussion of these notions. A dimension mapping on a Krull-Schmidt category K with Auslander-Reiten sequences is understood to be a mapping $\underline{\dim} : K \longrightarrow \bigoplus_I \mathbb{N}_0$, for some index set I, which is additive on short exact sequences. (Here, $\mathbb{N}_0$ denotes the set $\{0,1,2,\ldots\}$, and additivity of $\underline{\dim}$ on short exact sequences means that $\underline{\dim}\, X + \underline{\dim}\, Z = \underline{\dim}\, Y$, for all short exact sequences $X \rightarrowtail Y \twoheadrightarrow Z$ in K.)

A stable tube of rank r, $r \in \mathbb{N}$, is a translation quiver of the form $\mathbb{Z}\mathbb{A}_\infty/(\tau^r)$. If there is need to label the points of a stable tube of rank r then we always use $\mathbb{Z}/r\mathbb{Z} \times \mathbb{N}$ as index set, according to the following convention. We denote the points which belong to the mouth of the tube by t_{i1}, $i \in \mathbb{Z}/r\mathbb{Z}$, and we denote the points which belong to the ray starting in t_{i1} by t_{ij}, $j \in \mathbb{N}$.

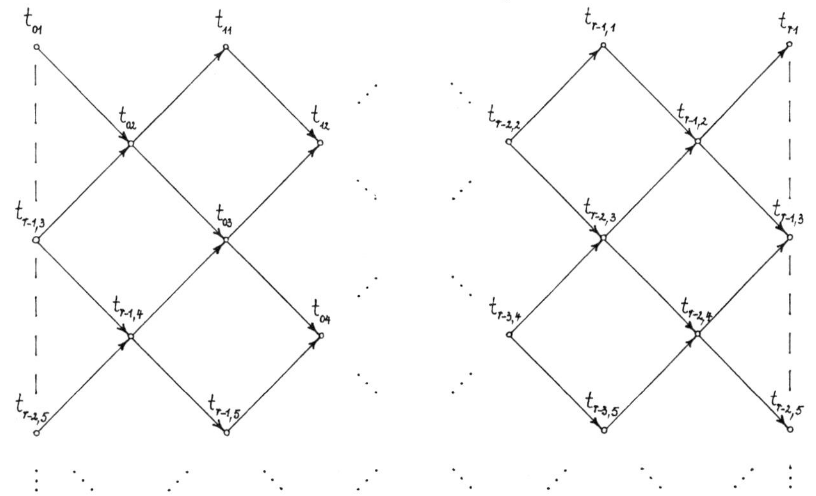

(Identify along the vertical interrupted lines.)

Now let K be a Krull-Schmidt category having Auslander-Reiten sequences, together with a fixed dimension mapping $\underline{\dim}\colon K \longrightarrow \bigoplus_I \mathbb{N}_0$. Following [Ri 84], we define a *tubular* I-*series* of K to be an I-series $T = \dot{\bigcup}_{\lambda\in I} T(\lambda)$ of components of $A_s(K)$, satisfying the following two conditions:

a) $T(\lambda)$ is a stable tube of rank $r(\lambda)$, for each $\lambda \in I$.

b) There exists a vector $\underline{d}(T) \in \bigoplus_I \mathbb{N}_0$ such that $\underline{\dim}\, t_{i,r(\lambda)}(\lambda) = \underline{d}(T)\deg(\lambda)$, for all $\lambda \in I$ subject to $T(\lambda) = \overline{T(\lambda)}$, and for all $i \in \mathbb{Z}/r(\lambda)\mathbb{Z}$. (In condition (b) we denote by $t_{ij}(\lambda)$ the points in $T(\lambda)$, in accordance with the above convention, by $\overline{T(\lambda)}$ the component of $A(K)$ containing $T(\lambda)$, and by $\deg : I \longrightarrow \mathbb{N}$ the function which is given by $\deg(\lambda) =$ degree of λ, for all $\lambda \in I\backslash\{\infty\}$, and $\deg(\infty) = 1$.)

In a tubular I-series $T = \dot{\bigcup}_{\lambda\in I} T(\lambda)$, the stable tubes of rank 1 are called the *homogeneous tubes*, and the stable tubes of rank > 1 are called the *exceptional tubes*. Also, the set of parameters $\lambda \in I$ for which $T(\lambda)$ is an exceptional tube, is called the set of *exceptional parameters* of T. Of course, the set of exceptional parameters of a tubular I-series T depends on the chosen parametrization of T by I.

With any tubular I-series T of K we associate its *tubular type* and its *dimension type*, a combinatorial invariant and an arithmetical invariant, as follows. Consider the I-family of valued graphs $\{\Gamma(\lambda)\}_{\lambda\in I}$, given by

$$\Gamma(\lambda) : \quad \underset{1}{\circ}\longrightarrow\!\!\!\!-\underset{2}{\circ}\!-\!- \;\cdots\; -\underset{r(\lambda)-1}{\circ}\overset{(1,\deg(\lambda))}{-\!\!-\!\!-\!\!-\!\!-}\underset{r(\lambda)}{\circ} ,$$

and define the tubular type of T to be the valued graph which one obtains from $\dot{\bigcup}_{\lambda\in I} \Gamma(\lambda)$ after identifying all points labelled by 1. In particular, the underlying graph of the tubular type of a tubular I-series is always a star. On the other hand, we define the dimension type of T to be $\underline{\dim}\, T = \underline{d}(T)$. Note that both, the tubular type and the dimension type of T are independent of the chosen parametrization of T by I .

We recur to the problem posed at the beginning, namely the investigation of the particular Krull-Schmidt category ${}_\Lambda L$. It is well--known that ${}_\Lambda L$ has Auslander-Reiten sequences ([Au 75],[RoS 76]). We set $A(\Lambda) = A({}_\Lambda L)$ and $A_s(\Lambda) = A_s({}_\Lambda L)$, for brevity. Here, the stable Auslander-Reiten quiver $A_s(\Lambda)$ is the full subquiver of $A(\Lambda)$ whose set of points is given by $\{[X] \mid X \in \mathrm{ind}_\Lambda L\ ,\ X \ncong \Lambda\}$. As fixed dimension mapping $\underline{\dim} : {}_\Lambda\mathcal{L} \longrightarrow \mathbb{N}_0$ we choose the R-rank: $\underline{\dim}\, M = \mathrm{rank}_R M$, for all $M \in {}_\Lambda L$.

In this set-up the solution of problem (II) can be formulated as follows:

THEOREM II. (i) The stable Auslander-Reiten quiver of Λ is given by a $\mathbb{P}_1\mathbb{Q}$-family of tubular I-series: $A_s(\Lambda) \cong \dot{\bigcup}_{\beta:\alpha\in\mathbb{P}_1\mathbb{Q}} T_{\beta:\alpha}$.

(ii) The tubular type of $T_{\beta:\alpha}$ does not depend on $\beta:\alpha \in \mathbb{P}_1\mathbb{Q}$. It only depends on K being a splitting field for C_3 or not. If K is a splitting field for C_3 then the tubular type is $\tilde{\mathbb{D}}_4$. If K is not a splitting field for C_3 then the tubular type is $\tilde{\mathbb{CD}}_3$.

(iii) For all $\beta:\alpha \in \mathbb{P}_1\mathbb{Q}$, the dimension type of $T_{\beta:\alpha}$ is given by

$$\underline{\dim}\, T_{\beta:\alpha} = |\beta|\ \underline{\dim}\, T_{1:0} + |\alpha|\ \underline{\dim}\, T_{0:1} = 6|\beta| + 3|\alpha| ,$$

where (β,α) is a pair of relatively prime integers representing $\beta:\alpha$.

(iv) The position of the unique projective-injective point $[\Lambda]$ of $A(\Lambda)$ is at the mouth of the exceptional tube $T_{0:1}(\infty)$:

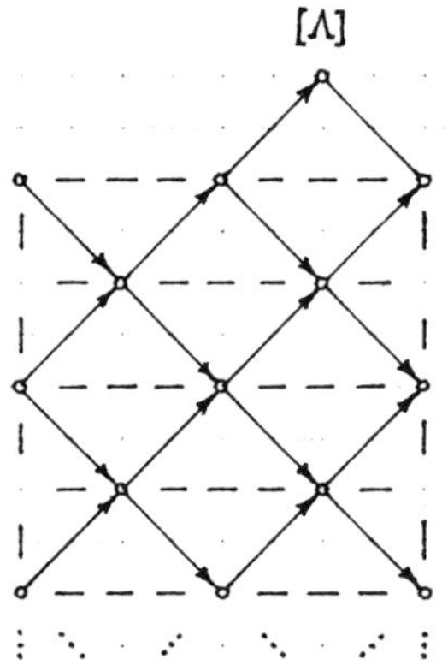

(The horizontal interrupted lines indicate the Auslander-Reiten translation.)

Choosing a fixed parametrization of $T_{\beta:\alpha}$ by I such that the set of exceptional parameters of $T_{\beta:\alpha}$ coincides with E, for all $\beta:\alpha \in \mathbb{P}_1\mathbb{Q}$, it becomes apparent that Theorem I is an immediate consequence of Theorem II. Therefore we only have to prove Theorem II.

Here, the two cases where K is a splitting field for C_3, respectively where K is not a splitting field for C_3, have to be treated separately. However, the proofs of Theorem II in these two cases differ only with respect to technical details, whereas their structure is completely analogous. Therefore we shall present the proof of Theorem II only for one of the possible situations, namely for the case where K is <u>not</u> a splitting field for C_3.

The general idea of the proof is to construct a sequence of Krull--Schmidt categories ${}_\Lambda L = L_0, L_1, L_2, L_3$ having the following properties:

(a) L_i is defined over the ground ring $R/\pi^{n_i}R$, for all $i = 0,1,2,3$ and $\infty = n_0 > n_1 > n_2 > n_3 = 1$.

(b) The Auslander-Reiten quivers $A(L_i)$ and $A(L_{i+1})$ are so closely related that their difference can be controlled, for all $i = 0,1,2$.

(c) There is a finite-dimensional tame k-algebra A of known Auslander-Reiten structure and a full subcategory of modA which is representation equivalent to L_3 , such that $A(L_3)$ can be derived from $A(\mathrm{A})$.

All passages from L_i to L_{i+1} are induced by functors which, however, are of different kinds and often seem to be strongly adapted to the particular situation. Therefore the entire passage from L_0 to L_3 is, strictly speaking, not of an algorithmic nature. Nevertheless the categories L_1, L_2, L_3 are of a significant common feature. Namely they all are full subcategories of generalized factorspace categories, a notion which we recall in section 0.2. In the text we shall use a different notation: $L_1 = F(K_0)$, $L_2 = F(K_1)$, $L_3 = \hat{C} \subset F(K)$.

In section 1 (first reduction) we reduce the length of the ground ring from ∞ to 3 , proceeding in two steps. In the first step we apply the main result from [Di 85] to the category ${}_\Lambda L$, thus obtaining a generalized factorspace category $F(K_0)$ which is defined over R/π^4R , such that $A(\Lambda) \cong A(F(K_0))$. In the second step we split off the unique indecomposable projective-injective object from $F(K_0)$, thus passing to a generalized factorspace category $F(K_1)$ which is defined over R/π^3R , such that $A_s(\Lambda) \cong A(F(K_1))$.

Sections 2,3 and 4 (second, third and fourth reduction) deal with the problem of reducing the length of the ground ring from 3 to 1. This reduction reveals a completely different nature and is much more difficult to handle than the first reduction. (Quite generally, in reducing the classification problem for the lattice category over any R-order to the classification problem for a related k-additive Krull-Schmidt category, it seems to be a significant feature that the first part, leading to a category defined over a ground ring of finite length, is rather accessible, whereas the second part, leading to a category defined over the residue class field, may be very involved.) We shall relate the generalized factorspace category $F(K_1)$ to the category $[F(K)]$ which is a quotient category of the (ordinary) factorspace category $F(K)$, defined over k. In order to do so we pass from $F(K_1)$ to its base--dependent version, the matrix category $\mathcal{M}$. The second reduction is a matrix reduction for $\mathcal{M}$ which, although becoming more and more involved, eventually leads to a large full subcategory $\hat{S}$ of $\mathcal{M}$. On the one hand, the matrix reduction gives an explicit description of the complement $T(E)$ of $\hat{S}$ in $\mathcal{M}$, and indeed, the third reduction establishes that the isomorphism classes of $\text{ind}T(E)$ are precisely the points of the three exceptional tubes $T_{1:1}(\infty)$, $T_{1:1}(0)$ and $T_{1:1}(\delta)$ of the tubular I-series $T_{1:1}$. Hence $\hat{S}$ is closed under Auslander--Reiten components and is "component-cofinite" in $\mathcal{M}$. On the other hand, all objects in $\hat{S}$, being of a partially reduced form, are matrices of a certain block structure. It is due to this block structure that we can find a functor $\phi : \hat{S} \longrightarrow [F(K)]$.

The fourth reduction deals with definition and analysis of the functor ϕ. This seems to be the most involved part of the whole proof. Note the difference to the second reduction, where we work with a matrix problem, that is, with objects and <u>chosen isomorphisms</u> of $\mathcal{M}$. In contrast, in the fourth reduction we investigate a functor, that is we have to work with objects and <u>all morphisms</u> of $\hat{S}$. The analysis of all morphisms soon leads to the analysis of a system of 435 quadratic congruences of matrix blocks modulo π^3 . Yet another difficulty arises from the fact that the non-reduced parts of matrices in $\hat{S}$ are of an angular form. Therefore, if we want to relate $\hat{S}$ to a factorspace category then we cannot define ϕ in the naive way, by assigning to a matrix in $\hat{S}$ just its non-reduced part. Instead, we have to adjoin a matrix of a prescribed normal form such that the now enlarged matrix can be interpreted as an object in $[F(K)]$. The third main difficulty lies in the fact that the full subcategory $[\hat{C}]$ of $[F(K)]$, onto which ϕ is dense, is only a subquotient category of the factorspace category $F(K)$, but is not a factorspace category itself. Notwithstanding these obstructions, working through quite formidable calculations we are able to prove that $\phi : \hat{S} \longrightarrow [\hat{C}]$ is a representation equivalence. Section 4 is completed by an elementary proof of the isomorphism $A([\hat{C}]) \cong A(\hat{C})$.

So far the problem of determining the Auslander-Reiten quiver of Λ^L has been reduced to the problem of determining the Auslander-Reiten quiver of $\hat{C}$, where $\hat{C}$ is defined over k . In section 5 we solve this latter problem by relating $\hat{C}$ to the module category modA of a tubular one-point extension A of a tame hereditary algebra of extended Dynkin type $\tilde{\mathbb{D}}_3$. The Auslander-Reiten structure of such tubular one-point extensions is known [Ri 84]. In this procedure we do not encounter any serious difficulties any more. In a first step we pass from $\hat{C}$ to its dual category $\hat{D}$, which is a full subcategory of the subspace category $U(\tilde{K})$. In a second step we introduce the one-point extension $A = A_0[R]$ of the tame hereditary algebra A_0 with respect to the simple regular A_0-module R , such that $\mathrm{Hom}_{A_0}(R, \mathrm{preinj}A_0)$ is isomorphic to $\tilde{K}$. We complete the proof by making use of the well-known connection between the categories modA and $U(\mathrm{Hom}(R, \mathrm{preinj}A_0))$.

There will be some material in the proof which is not asserted in Theorem I or Theorem II. This will be recorded in section 6 (appendix). It is concerned with the effective constructibility of the bijection between $\mathbb{P}_1\mathbb{Q} \times \bar{I} \times \mathbb{N}$ and the set of all isomorphism classes of nonprojective indecomposable Λ-lattices, and with the Auslander-Reiten quiver of the category $\mathcal{M}$. It also contains a Leitfaden which shows the structure of the proof in terms of a diagram of categories and functors, and which the reader may consult whenever he needs orientation.

In a preliminary section 0 we fix notations and conventions, and we define once and for all several Krull-Schmidt categories which play an important rôle throughout the proof. Moreover, for a few finite dimensional k-algebras of finite or tame representation type we fix normal forms for their indecomposable modules. We shall need them in the second reduction, where these k-algebras occur as local problems of the matrix problem.

It seems appropriate to draw the essence from the entire reduction process sketched above. We encounter the fact that all four reductions are qualitatively fundamentally different from each other. We emphasize that this feature seems to be most significant for the whole proof, reflecting its complexity to some extent. We summarize (see section 0 for definition of symbols):

1. The first reduction reduces the problem of determining $A(\Lambda)$ to the problem of determining $A(F(K_1))$. The approach is to investigate Ext--groups.

2. The second reduction reduces the problem of determining $A(F(K_1))$ to the problem of determining $A_{\mathcal{M}}(\hat{S} \vee T(E))$. The approach is to reduce matrices.

3. The third reduction reduces the problem of determining $A_{\mathcal{M}}(\hat{S} \vee T(E))$.
to the problem of determining $A(\hat{S})$. The approach is to construct Auslander-Reiten components.

4. The fourth reduction reduces the problem of determining $A(\hat{S})$ to the problem of determining $A(\hat{C})$. The approach is to construct and analyse a functor.

5. The problem of determining $A(\hat{C})$ is solved by application of known results from the theory of tame algebras.

0. PRELIMINARIES.

All material presented in this section (notation, conventions, definitions, fixed normal forms etc.) is obligatory for all the subsequent text. It will remain unchanged and valid throughout.

0.1. NOTATION AND CONVENTIONS.

For sets of numbers we use the following notation: $\mathbb{N} = \{1,2,3,\ldots\}$, $\mathbb{N}_0 = \mathbb{N} \cup \{0\}$, $\mathbb{N}_\infty = \mathbb{N} \cup \{\infty\}$, $\mathbb{Q}^+ = \{r \in \mathbb{Q} \mid r > 0\}$, $\mathbb{Q}_0^+ = \mathbb{Q}^+ \cup \{0\}$, $\mathbb{Q}_{0,\infty}^+ = \mathbb{Q}^+ \cup \{0,\infty\}$.

Throughout, R denotes a complete discrete valuation ring with exponential valuation v , satisfying the following properties:

(a) $v(3) = 4$,

(b) R does not contain a primitive third root of unity.

With R we associate the following data:

K = field of fractions of R

k = residue class field of R

π = chosen parameter of R

$R_n = R/\pi^n R \quad (n \in \mathbb{N}_\infty)$

$d = \frac{-3}{\pi^4}$, unit of R by hypothesis (a)

$\bar{d}$ = residue class of d in k

$\delta = X^2 - \bar{d}$, irreducible in $k[X]$ by hypothesis (b)

$f = k[X]/(\delta)$

$I = \{\lambda \in k[X] \mid \lambda$ is monic and irreducible$\} \cup \{\infty\}$

$E = \{\infty, 0, \delta\}$

$\check{I} = I \setminus E$.

Throughout, $\Lambda = RC_3$ denotes the group ring which is given by the cyclic group of order 3, and ${}_\Lambda L$ denotes the category of all left-Λ-lattices. More generally, if Ξ is any R-order then ${}_\Xi L$ denotes the category of all left Ξ-lattices.

For any ring S, modS denotes the category of all finitely generated left S-modules. Quite in general, all modules considered will be finitely generated and will be left modules. Composition of morphisms will always be written on the left.

Let K be a Krull-Schmidt category. We shall use the following notation:

X^n = direct sum of n copies of X, for any object $X \in K$ and $n \in \mathbb{N}$.

$K(X,Y)$ = set of all K-morphisms form X to Y, for any two objects $X,Y \in K$.

ind$K$ = full subcategory of K consisting of all indecomposable objects of K .

Starting from full subcategories of K, we devote the symbol $\vee$ to the formation of the new full subcategory of K which is generated by these with respect to finite direct sums. More precisely, let $\{K_i\}_{i\in I}$ be any family of full subcategories of K . Then we denote by $\bigvee_{i\in I} K_i$ the full subcategory of K whose class of objects is given by $\{\oplus_{i\in I'} X_i \mid I' \subset I,\ I' \text{ finite},\ X_i \in K_i\}$. We emphasize two special cases of this formation:

1) In case $I = \{1,2\}$ we write $\bigvee_{i\in\{1,2\}} K_i = K_1 \vee K_2$.

2) In case all subcategories K_i consist only of a single element

X_i we write $\bigvee_{i\in I} K_i = \mathrm{add}\{X_i \mid i \in I\}$.

Let K be a Krull-Schmidt category and let K' be a full subcategory of K . We associate with K and K' the following quivers:

$A(K)$ = Auslander-Reiten quiver of K ,

$A(K')$ = Auslander-Reiten quiver of K' ,

$A_K(K')$ = full subquiver of $A(K)$ whose set of points is given by the isomorphism classes of indecomposable objects in K' .

(Note that for all $X, Y \in \mathrm{ind}K'$ there is a canonical bimodule epimorphism $\mathrm{irr}_{K'}(X,Y) \twoheadrightarrow \mathrm{irr}_K(X,Y)$. Therefore the sets of points of the quivers $A(K')$ and $A_K(K')$ coincide, and if there is an arrow $[X] \xrightarrow{(b,b')} [Y]$ in $A_K(K')$ then there is an arrow $[X] \xrightarrow{(a,a')} [Y]$ in $A(K')$, subject to $a \geq b$ and $a' \geq b'$. In the text we shall be particularly interested in subcategories K' of K for which $A_K(K') = A(K')$.)

The term quiver-isomorphism or isomorphism of quivers does not necessarily mean isomorphism of translation quivers. Indeed, most of the asserted quiver-isomorphisms in the text will not be isomorphisms of translation quivers. Note however, that the quiver-isomorphism in Theorem II (i) in fact is an isomorphism of translation quivers because on a stable tube there exists only one translation mapping.

A functor $\phi : K \longrightarrow K'$ between Krull-Schmidt categories K and K' is said to be R-additive, provided all mappings $\phi_{X,Y} : K(X,Y) \longrightarrow K'(\phi(X),\phi(Y))$, which are given by ϕ for any two objects $X, Y \in K$, are R-linear mappings. An R-additive functor $\phi : K \longrightarrow K'$ between Krull-Schmidt categories is said to be a representation equivalence, provided it is dense, full and isomorphism-reflecting. (Note that a representation equivalence $\phi : K \longrightarrow K'$ induces a bijection between the sets of points of $A(K)$ and $A(K')$. Moreover, it induces an isomorphism of quivers $A(K) \cong A(K')$ if and only if $\ker\phi_{X,Y} \subset \mathrm{rad}^2_K(X,Y)$, for all $X,Y \in \mathrm{ind}K$.)

For any commutative ring S and any pair $(m,n) \in \mathbb{N}_0 \times \mathbb{N}_0$ we denote by $S^{m\times n}$ the set of all $m\times n$ - matrices over S . (In the text S will be one of the rings R_n , for some $n \in \mathbb{N}$.) Some special types of matrices will be denoted by special symbols:

E_n = identity matrix in $S^{n\times n}$ (Usually, we simply write $E = E_n$.)

I = unique matrix in $S^{1\times 0}$

$\vdash\!\dashv$ = unique matrix in $S^{0\times 1}$

ϕ_{λ^n} = Frobenius matrix of λ^n in $k^{m\times m}$, $m = n\cdot\deg\lambda$, for all $\lambda \in I\setminus\{\infty\}$, $n \in \mathbb{N}$.

In matrices with block structure, blocks carrying no symbol are understood to be zero blocks. For any matrix M , M^t denotes its transposed matrix. If $\varphi : X \longrightarrow Y$ is a morphism in modR then we denote by (φ) the matrix corresponding to φ with respect to chosen bases in X and Y . Throughout, matrices will be viewed within the column calculus. That is to say, vectors correspond to columns and linear maps correspond to sets of columns. (Note that this convention fits to the above agreement on the composition of morphisms: if $\varphi : X \longrightarrow Y$ and $\psi : Y \longrightarrow Z$ are morphisms in modR then $(\psi\varphi) = (\psi)(\varphi)$, for any fixed choice of bases of X,Y,Z .)

0.2. GENERALIZED FACTORSPACE CATEGORIES.

Let $(K,|\cdot|)$ be a Krull-Schmidt category K together with an R-additive functor $|\cdot| : K \longrightarrow \mathrm{mod}R_n$, for some $n \in \mathbb{N}_\infty$. With the pair $(K,|\cdot|)$ we associate the <u>generalized factorspace category</u>

$\mathcal{F}=\mathcal{F}(\mathcal{K},|\cdot|)$. Objects in $\mathcal{F}$ are triples (K,F,φ) , where $K \in \mathcal{K}$, F is a finitely generated injective R_n-module, and $\varphi \in \mathrm{Hom}_{R_n}(|K|,F)$. Morphisms in $\mathcal{F}$ from (K,F,φ) to (K',F',φ') are pairs $(|\beta|,\alpha)$, where $\beta \in \mathcal{K}(K,K')$, $\alpha \in \mathrm{Hom}_{R_n}(F,F')$, and $\alpha\varphi = \varphi'|\beta|$.

Note the special case $n = 1$. In that case $(\mathcal{K},|\cdot|)$ is a vectorspace category over k and $\mathcal{F}(\mathcal{K},|\cdot|)$ is the factorspace category (in the ordinary sense) of the vectorspace category $(\mathcal{K},|\cdot|)$.

In our context the functor $|\cdot| : \mathcal{K} \longrightarrow \mathrm{mod}R_n$ usually will be an embedding. In that case we omit the symbol $|\cdot|$, writing $\mathcal{F}(\mathcal{K}) = \mathcal{F}(\mathcal{K},|\cdot|)$ and $(\beta,\alpha) = (|\beta|,\alpha)$.

Now let $\mathcal{F}(\mathcal{K})$ be the generalized factorspace category of a Krull-Schmidt subcategory $\mathcal{K}$ of $\mathrm{mod}R_n$. We clarify the notions short exact sequence in $\mathcal{F}(\mathcal{K})$, Auslander-Reiten sequence in $\mathcal{F}(\mathcal{K})$ and pattern of $\mathcal{K}$. A sequence of two morphisms $(K_1,F_1,\varphi_1)\xrightarrow{(\beta_1,\alpha_1)}(K_2,F_2,\varphi_2)\xrightarrow{(\beta_2,\alpha_2)}(K_3,F_3,\varphi_3)$ in $\mathcal{F}(\mathcal{K})$ is said to be a <u>short exact sequence</u> in $\mathcal{F}(\mathcal{K})$, provided both sequences $K_1\xrightarrow{\beta_1}K_2\xrightarrow{\beta_2}K_3$ and $F_1\xrightarrow{\alpha_1}F_2\xrightarrow{\alpha_2}F_3$ are split short exact sequences in $\mathrm{mod}R_n$. A short exact sequence $X_1\xrightarrow{\xi_1}X_2\xrightarrow{\xi_2}X_3$ in $\mathcal{F}(\mathcal{K})$ is called an <u>Auslander-Reiten sequence</u> in $\mathcal{F}(\mathcal{K})$ if it satisfies the following conditions:

(a_1) The sequence does not split.

(a_2) X_1 and X_3 are indecomposable.

(a_3) For each morphism $\eta : X_1 \longrightarrow Y$ which is not a splittable monomorphism there exists a morphism $\eta' : X_2 \longrightarrow Y$ such that $\eta = \eta'\xi_1$.

(a_4) For each morphism $\zeta : Z \longrightarrow X_3$ which is not a splittable epimorphism there exists a morphism $\zeta' : Z \longrightarrow X_2$ such that $\zeta = \xi_2\zeta'$.

The <u>pattern</u> of K is, by definition, the diagram in $\mathrm{mod}R_n$ which is given by a chosen set of representatives for the isomorphism classes of indecomposable objects of K, together with a chosen system of generating morphisms of K. (This is related to, but not identical with the original meaning of the notion "pattern", as introduced by Ringel in [Ri 79].) Of course, $F(K)$ is uniquely determined by the pattern of K. For indicating the objects of patterns, sometimes together with their endomorphism rings, we use the following set of symbols:

 $= R_4$, 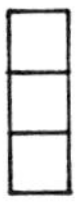 $= R_3$, 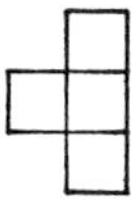 $= R_1 \oplus R_3$, 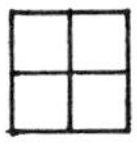 $= R_2 \oplus R_2$,

$\bullet = k$,

$\blacksquare = k \oplus k$, and $K(\blacksquare,\blacksquare) = \left\{ \begin{pmatrix} a & 0 \\ 0 & a \end{pmatrix} \;\middle|\; a \in k \right\} \cong k$.

$\CIRCLE = k \oplus k$, and $K(\CIRCLE,\CIRCLE) = \left\{ \begin{pmatrix} a & b\bar{d} \\ b & a \end{pmatrix} \;\middle|\; a, b \in k \right\} \cong f$.

To each morphism $\mu : K_i \longrightarrow K_j$ of a pattern we attach the matrix which corresponds to μ with respect to the canonical bases of K_i and K_j .

We shall also need the concept of a <u>subspace category</u>. Suppose $n = 1$ and let $(K, |\cdot|)$ be a vectorspace category over k . Then the subspace category $U = U(K, |\cdot|)$ of the vectorspace category $(K, |\cdot|)$ is defined dually to the factorspace category $F(K, |\cdot|)$. In case the functor $|\cdot| : K \longrightarrow \mathrm{mod}k$ is an embedding, we write $U(K) = U(K, |\cdot|)$. The notions short exact sequence in $U(K)$ and Auslander-Reiten sequence in $U(K)$ are defined as for factorspace categories. We remark that $U(K)$ is uniquely determined by the pattern of K .

DEFINITION. (Generalized factorspace categories and a subspace category). Let $K_0 \subset \mathrm{mod}R_4$, $K_1 \subset \mathrm{mod}R_3$, $K \subset \mathrm{mod}k$, $\tilde{K} \subset \mathrm{mod}k$ be the Krull-Schmidt categories which are given by the following patterns.

K_0 :

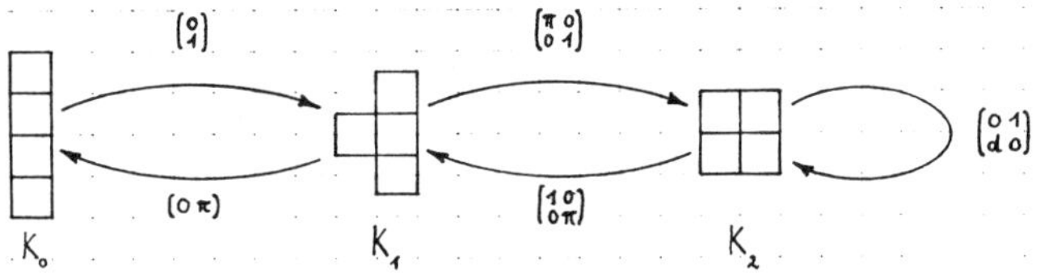

K_1 :

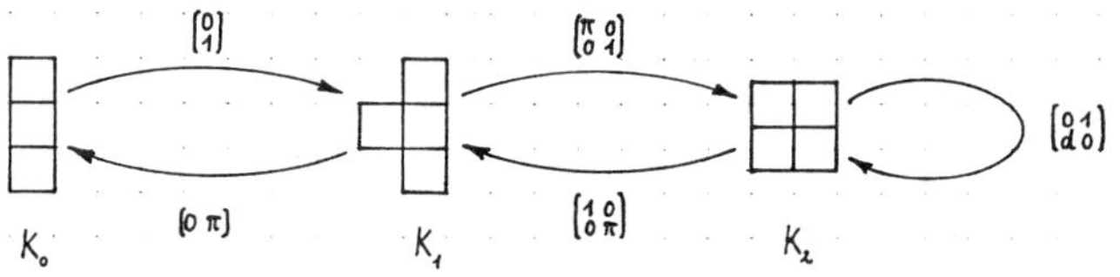

K :

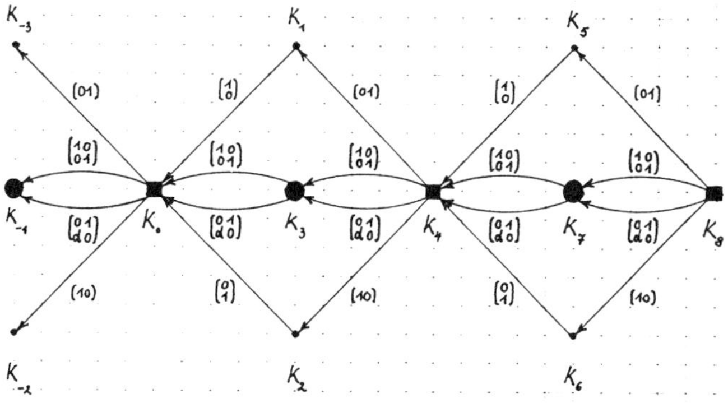

$\tilde{K}$:

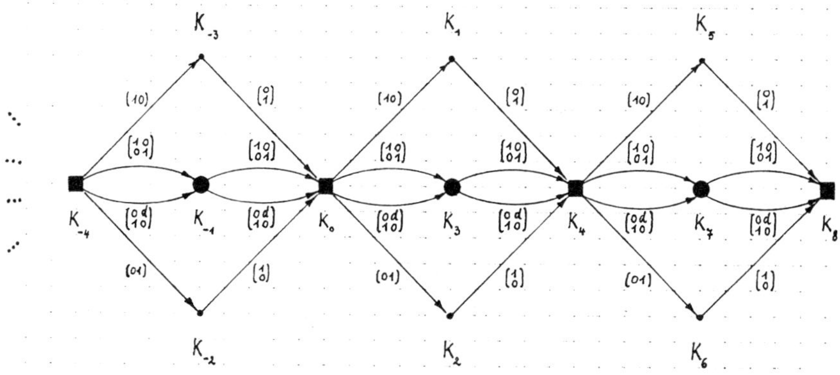

With these Krull-Schmidt categories we associate the generalized factorspace categories $F(K_0)$, $F(K_1)$, $F(K)$ and the subspace category $U(\tilde{K})$. The labelling of the indecomposable objects of K_0 , K_1 , K , $\tilde{K}$ by symbols K_i , as indicated in the patterns, will be fixed throughout. Moreover, for any pair of integers $a,b \in \mathbb{Z}$, subject to $a \leq b \leq 8$, we set

$K_{[a,b]}$ = full subcategory $\{\text{add}K_i \mid i \in [a,b]\}$ of K ,

$\tilde{K}_{[a,b]}$ = full subcategory $\{\text{add}K_i \mid i \in [a,b]\}$ of $\tilde{K}$.

0.3. NORMAL FORMS FOR LOCAL PROBLEMS.

In the first part of this subsection we introduce some notation which centers around the concept of a k-species. For details concerning the representation theory of k-species of finite and tame representation type the reader is referred to [Dl/Ri 76] and [Ri 84].

Let S be a k-species with given modulation. We set

kS = path algebra of S with respect to k ,

rep(k,S) = category of all k-representations of S ,

A(S) = $A(k\text{S}) = A(\text{rep}(k,\text{S}))$.

Freely we shall identify the categories rep(k,S) and mod$k$S .

Let S be a tame k-species of extended Dynkin type $\tilde{\Delta}$, with given modulation. We set

P = preprojective component of A(S) ,

R = $\dot{\bigcup}_{\lambda \in I} R(\lambda)$ = I-family of all regular components of A(S),

Q = preinjective component of A(S) ,

preproj$k$S = $\text{add}\{X \mid [X] \in P\}$,

$\text{reg}_\lambda k$S = $\text{add}\{X \mid [X] \in R(\lambda)\}$, for any $\lambda \in I$,

preinj$k$S = $\text{add}\{X \mid [X] \in Q\}$.

Note that the zero object is contained in each of the full subcategories preproj$k$S , reg$_\lambda k$S and preinj$k$S of mod$k$S . Recall that R is a tubular I-series of mod$k$S which is of tubular type Δ and which separates P from Q . Moreover, the Auslander-Reiten quiver of mod$k$S is given by $A(\mathrm{S}) = P \mathbin{\dot\cup} R \mathbin{\dot\cup} Q$.

Aiming at the definition of particular k-species and their modulation we agree that

 denotes an arrow with valuation (1,1) and modulation $k \overset{k}{\longleftarrow} k$,

 denotes an arrow with valuation (1,2) and modulation $k \overset{f}{\longleftarrow} f$,

 denotes an arrow with valuation (2,1) and modulation $f \overset{f}{\longleftarrow} k$.

DEFINITION (k-species). Let Q_1, Q_2, Q_3, S_1, S_2 and S_0 be the following k-species:

In the second reduction the k-species Q_1, Q_2, Q_3, S_1, S_2 will appear as "local problems". Preparatory to this we now recall their representation theory in terms of their Auslander-Reiten quiver, with points

replaced by normal forms of k-matrices. Throughout, these normal forms will be fixed. Hence in base-dependent terms, arbitrary k-representations of one of the species Q_1, Q_2, Q_3, S_1, S_2 will always be given as finite direct sums which are built from these fixed normal forms. In a k-representation of any of these species, we denote by A, A', A'' the k-matrices which correspond to the k-linear maps $\alpha, \alpha', \alpha''$ with respect to chosen bases.

1) <u>The category</u> $rep(k, Q_1)$. Objects are given by matrices (A) . The Auslander-Reiten quiver is given by

$A(Q_1)$:

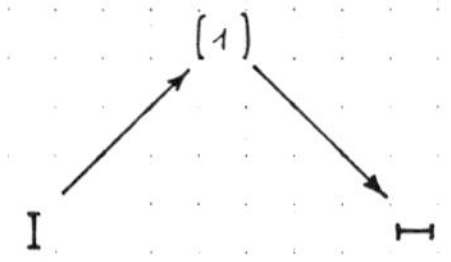

2) <u>The category</u> $rep(k, S_1)$. Objects are given by triples of matrices $(A \| A'_1 | A'_2)$. The Auslander-Reiten quiver is given by

$A(S_1)$:

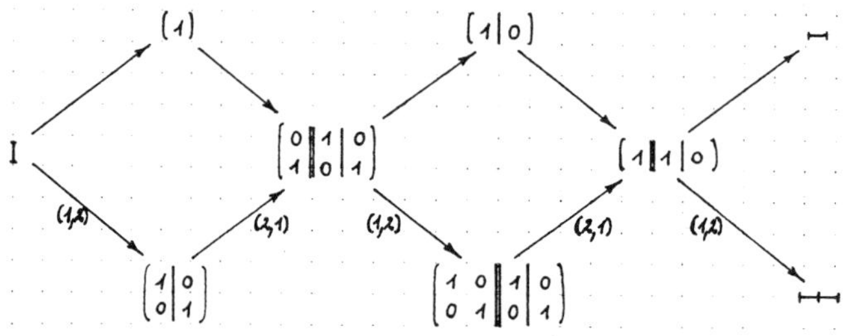

3) The category $rep(k, S\ )$. Objects are given by triples of matrices $\left(\begin{array}{c} A \\ \hline\hline A'_1 \\ \hline A'_2 \end{array}\right)$. The Auslander Reiten quiver $A(S_2)$ is obtained from $A(S_1)$ by reverting all arrows and transposing all matrices.

4) <u>The category</u> $\mathrm{rep}(k, Q_2)$. Objects are given by pairs of matrices $(A|A')$. A complete set of representatives for the isomorphism classes of indecomposable objects in $\mathrm{rep}(k, Q_2)$ is given by the following list.

a) The indecomposable preprojective representations:

$$(P_n \mid P'_n) = \left(\underbrace{\begin{matrix} 1 & & \\ & \ddots & \\ & & 1 \\ 0 & \cdots & 0 \end{matrix}}_{n} \left| \underbrace{\begin{matrix} 0 & \cdots & 0 \\ 1 & & \\ & \ddots & \\ & & 1 \end{matrix}}_{n} \right.\right) \qquad , \; n \in \mathbb{N}_0$$

b) The indecomposable regular representations:

$$(R(\lambda)_n \mid R(\lambda)'_n) = (E \mid \phi_\lambda n) \quad , \text{ if } \lambda \in I\setminus\{\infty\} \qquad , \; n \in \mathbb{N}$$

$$(R(\lambda)_n \mid R(\lambda)'_n) = (\phi_X n \mid E) \quad , \text{ if } \lambda = \infty \qquad , \; n \in \mathbb{N}$$

c) The indecomposable preinjective representations:

$$(I_n \mid I'_n) = \left.\left(\begin{matrix} 1 & & & 0 \\ & \ddots & & \vdots \\ & & 1 & 0 \end{matrix} \left| \begin{matrix} 0 & 1 & & \\ \vdots & & \ddots & \\ 0 & & & 1 \end{matrix}\right.\right)\right\} n \qquad , \; n \in \mathbb{N}_0$$

We denote by p_n, $r(\lambda)_n$, i_n the points in $A(Q_2)$ which correspond to these indecomposable representations. Then the components of $A(Q_2)$ are given as follows.

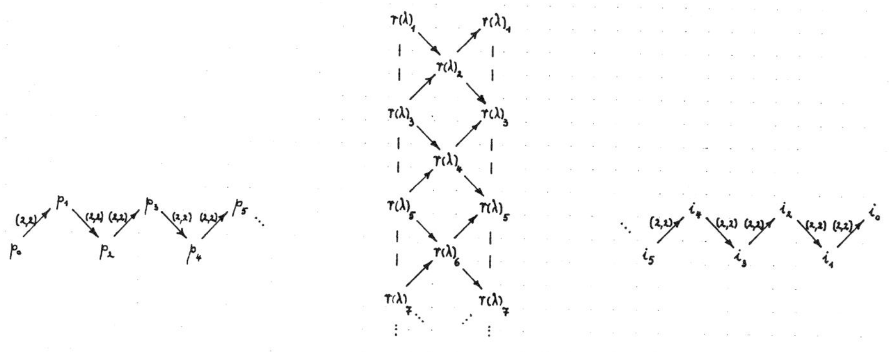

$\mathcal{P}$ $\qquad$ $\mathcal{R}(\lambda)$, $\lambda \in \mathcal{I}$ $\qquad$ $\mathcal{Q}$

5) <u>The category</u> $\operatorname{rep}(k,Q_3)$. Objects are given by triples of matrices $\begin{array}{|c|c}\hline A' & \\ \hline A & \multicolumn{1}{c|}{A''} \\ \hline\end{array}$. A complete set of representatives for the isomorphism classes of indecomposable objects in $\operatorname{rep}(k,Q_3)$ is given by the following list.

a) The indecomposable preprojective representations:

$$\begin{array}{|c|c}\hline P'_{2m} & \\ \hline P_{2m} & \multicolumn{1}{c|}{P''_{2m}} \\ \hline\end{array} = \begin{array}{|c|c}\hline E_m & \\ \hline \begin{matrix}1 & & \\ & \ddots & \\ & & 1\\ 0 & \cdots & 0\end{matrix} & \multicolumn{1}{c|}{\begin{matrix}0 & \cdots & 0\\ 1 & & \\ & \ddots & \\ & & 1\end{matrix}} \\ \hline\end{array} , \quad \begin{array}{|c|c}\hline P'_{2m+1} & \\ \hline P_{2m+1} & \multicolumn{1}{c|}{P''_{2m+1}} \\ \hline\end{array} = \begin{array}{|c|c}\hline \begin{matrix}0 & \cdots & 0\\ 1 & & \\ & \ddots & \\ & & 1\end{matrix} & \\ \hline \begin{matrix}1 & & \\ & \ddots & \\ & & 1\\ 0 & \cdots & 0\end{matrix} & \multicolumn{1}{c|}{E_{m+1}} \\ \hline\end{array} , \quad m \in \mathbb{N}_0 .$$

b) The indecomposable regular representations:

$$\begin{array}{|c|c}\hline R(\lambda)'_n & \\ \hline R(\lambda)_n & \multicolumn{1}{c|}{R(\lambda)''_n} \\ \hline\end{array} = \begin{array}{|c|c}\hline E & \\ \hline \Phi_{\lambda^n} & \multicolumn{1}{c|}{E} \\ \hline\end{array} , \quad \text{if } \lambda \in \mathcal{J}\setminus\{\infty\} , \quad n \in \mathbb{N} ;$$

$$\begin{array}{|c|c}\hline R(\infty)'_{2m} & \\ \hline R(\infty)_{2m} & \multicolumn{1}{c|}{R(\infty)''_{2m}} \\ \hline\end{array} = \begin{array}{|c|c}\hline E & \\ \hline E & \multicolumn{1}{c|}{\Phi_{X^m}} \\ \hline\end{array} , \quad \begin{array}{|c|c}\hline R(\bar\infty)'_{2m} & \\ \hline R(\bar\infty)_{2m} & \multicolumn{1}{c|}{R(\bar\infty)''_{2m}} \\ \hline\end{array} = \begin{array}{|c|c}\hline \Phi_{X^m} & \\ \hline E & \multicolumn{1}{c|}{E} \\ \hline\end{array} , \quad m \in \mathbb{N} ;$$

$$\begin{array}{|c|c}\hline R(\infty)'_{2m-1} & \\ \hline R(\infty)_{2m-1} & \multicolumn{1}{c|}{R(\infty)''_{2m-1}} \\ \hline\end{array} = \begin{array}{|c|c}\hline \begin{matrix}1 & & \\ & \ddots & \\ & & 1\\ 0 & \cdots & 0\end{matrix} & \\ \hline E_{m-1} & \multicolumn{1}{c|}{\begin{matrix}0 & 1 & & \\ \vdots & & \ddots & \\ 0 & & & 1\end{matrix}} \\ \hline\end{array} , \quad \begin{array}{|c|c}\hline R(\bar\infty)'_{2m-1} & \\ \hline R(\bar\infty)_{2m-1} & \multicolumn{1}{c|}{R(\bar\infty)''_{2m-1}} \\ \hline\end{array} = \begin{array}{|c|c}\hline \begin{matrix}1 & & & 0\\ & \ddots & & \vdots\\ & & 1 & 0\end{matrix} & \\ \hline E_m & \multicolumn{1}{c|}{\begin{matrix}0 & \cdots & 0\\ 1 & & \\ & \ddots & \\ & & 1\end{matrix}} \\ \hline\end{array} , \quad m \in \mathbb{N} .$$

c) The indecomposable preinjective representations:

$$\begin{array}{|c|c} \hline I'_{2m} & \\ \hline I_{2m} & I''_{2m} \\ \hline \end{array} = \begin{array}{|c|c} \hline \begin{matrix} 0 & 1 & & \\ \vdots & & \ddots & \\ 0 & & & 1 \end{matrix} & \\ \hline \begin{matrix} 1 & & & 0 \\ & \ddots & & \vdots \\ & & 1 & 0 \end{matrix} & E_m \\ \hline \end{array} \; , \quad \begin{array}{|c|c} \hline I'_{2m+1} & \\ \hline I_{2m+1} & I''_{2m+1} \\ \hline \end{array} = \begin{array}{|c|c} \hline E_{m+1} & \\ \hline \begin{matrix} 1 & & & 0 \\ & \ddots & & \vdots \\ & & 1 & 0 \end{matrix} & \begin{matrix} 0 & 1 & & \\ \vdots & & \ddots & \\ 0 & & & 1 \end{matrix} \\ \hline \end{array} \; , \quad m \in \mathbb{N}_0 \, .$$

We denote by p_n, $r(\lambda)_n$ (with $\lambda \in I \cup \{\bar{\infty}\}$) , i_n the points in $A(Q_3)$ which correspond to these indecomposable representations. Then the components of $A(Q_3)$ are given on the following page.

Note that by giving normal forms for the indecomposable objects of $\mathrm{rep}(k,Q_2)$ and $\mathrm{rep}(k,Q_3)$ we also have fixed a parametrization for the family of regular components of $A(Q_2)$ and $A(Q_3)$ by I .

The tame k-species S_0 will appear no sooner than in section 5. As to $A(S_0)$, we shall use the structure of its preinjective component, the dimension types of its preinjective points, and normal forms only for the five points which constitute the last preinjective mesh. All of this will be presented in section 5, when needed.

0.4 ANGULAR MATRICES AND DEFINITION OF SUBCATEGORIES.

In subsection 0.2 we have introduced the Krull-Schmidt categories $F(K_1)$, $F(K)$, $U(\tilde{K})$. This subsection is devoted to the definition of distinguished full subcategories $S \subset \hat{S} \subset F(K_1)$, $\hat{C} \subset F(K)$, $\hat{D} \subset U(\tilde{K})$, which will be of great importance for the solution of our problem. Their definition is based upon the technical device of angular matrices, a notion which will play a key rôle throughout. We begin with its definition.

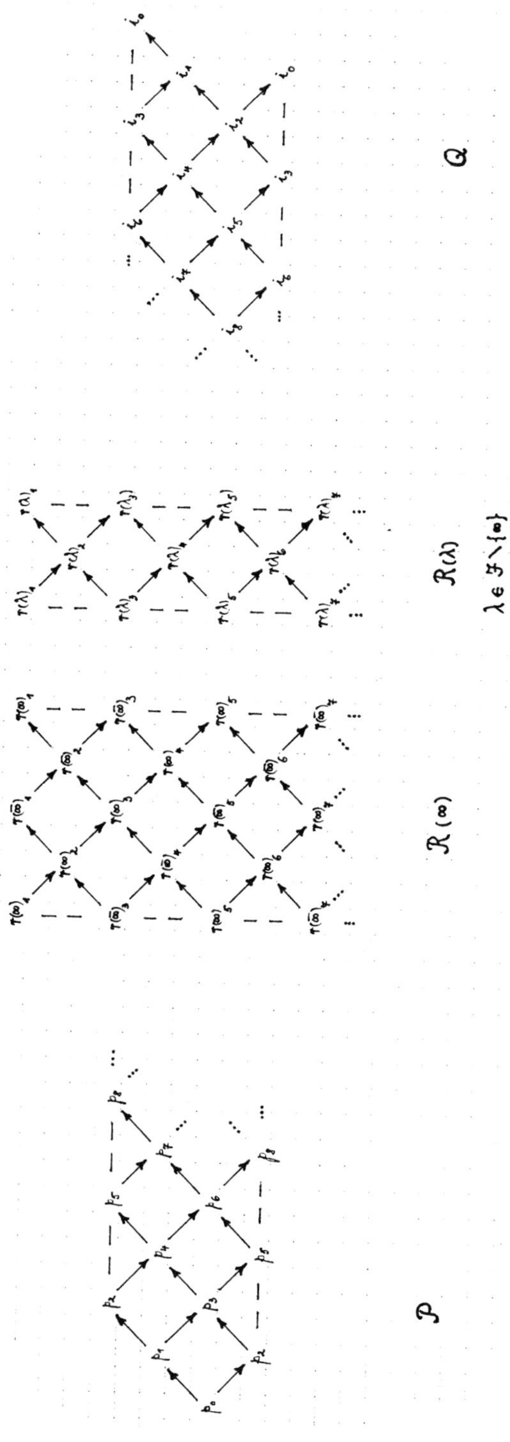
Q
R(λ)
λ ∈ F \ {∞}
R(∞)
P

DEFINITION (Angular matrices). Let $I_X = \{0,0',1,2,3,3',4,4',5,6,7,7',8,8'\}$, $I_Y = \{0,0',1,2,3,3',4,4'\}$, $I_Z = \{1,2,3\}$ be fixed index sets. An angular matrix (X,Y,Z) is, by definition, a set of R_3-matrices $(X,Y,Z) = \{X_i, Y_j, Z_k \mid i \in I_X,\ j \in I_Y,\ k \in I_Z\}$ which satisfy the following conditions:

(a_1) $X_i \in R_3^{m_0 \times n_i}$, $Y_j \in R_3^{m_j \times n_0}$, $Z_1 \in R_3^{m_1 \times n_1}$, $Z_2 \in R_3^{m_2 \times n_2}$,

$$Z_3 = \left(\begin{array}{c|c} Z_{3'3} & Z_{3'3'} \\ \hline Z_{33} & Z_{33'} \end{array}\right) \in R_3^{2m_3 \times 2n_3}, \text{ where } \{n_i, m_j \mid i \in I_X,\ j \in I_Y\} \subset \mathbb{N}_0 .$$

(a_2) $n_i = n_{i'}$, for all $i = 0,3,4,7,8$,

$m_j = m_{j'}$, for all $j = 0,3,4$.

(a_3) $X_0 = Y_0$ and $X_{0'} = Y_{0'}$.

In other words, the sizes of the matrices X_i, Y_j, Z_k which constitute an angular matrix (X,Y,Z) fit together to the following arrangement.

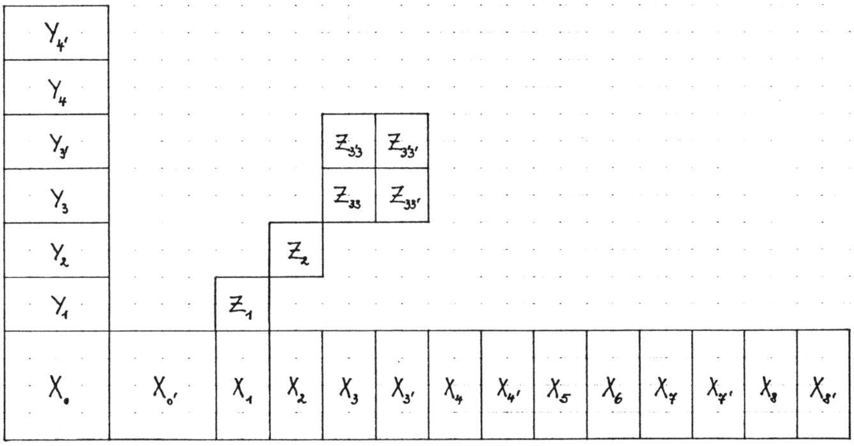

We denote by $\mathcal{A}$ the set of all angular matrices.

We proceed by introducing mappings $\lambda : \mathcal{A} \to \operatorname{rep}(k, Q_3)$, $\rho : \mathcal{A} \to \operatorname{rep}(k, Q_3)$, $\lambda' : \mathcal{A} \to U(\tilde{K}_{[0,4]})$, $\rho' : \mathcal{A} \to F(K_{[0,8]})$, which will serve for the definition of subsets $\mathfrak{M}$ and $\hat{\mathfrak{M}}$ of $\mathcal{A}$.

For any $(X,Y,Z) \in \mathcal{A}$, let $\eta_j : k^{n_0} \to k^{m_j}$ be the k-linear map which is given by Y_j modulo π, for all $j \in I_Y$. Set $\check{I}_Y = I_Y \setminus \{2,4,4'\}$, consider $\eta = (\eta_j)_{j\in\check{I}_Y} : k^{n_0} \longrightarrow \bigoplus_{j\in\check{I}_Y} k^{m_j}$, and denote by $\kappa : \ker\eta \longrightarrow k^{n_0}$ the corresponding kernel map. Then $\lambda(X,Y,Z)$ is, by definition, the following k-representation of Q_3 :

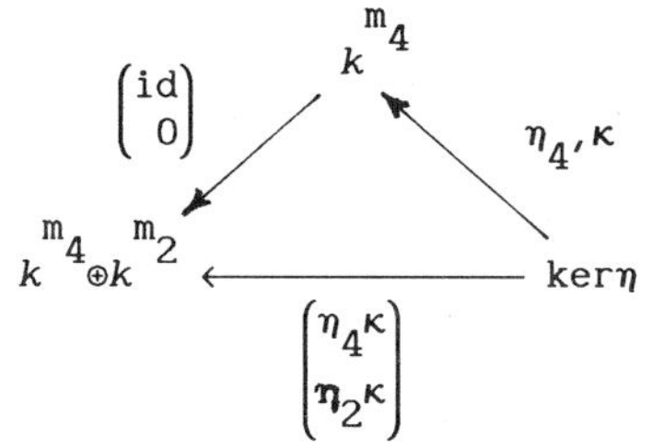

Also, for any $(X,Y,Z) \in \mathcal{A}$, let $\xi_i : k^{n_i} \longrightarrow k^{m_0}$ be the k-linear map which is given by X_i modulo π, for all $i \in I_X$. Set $\check{I}_X = I_X \setminus \{5,8,8'\}$, consider $\xi = (\xi_i)_{i\in\check{I}_X} : \bigoplus_{i\in\check{I}_X} k^{n_i} \longrightarrow k^{m_0}$, and denote by $\zeta : k^{m_0} \longrightarrow \operatorname{coker}\xi$ the corresponding cokernel map. Then $\rho(X,Y,Z)$ is, by definition, the following k-representation of Q_3 :

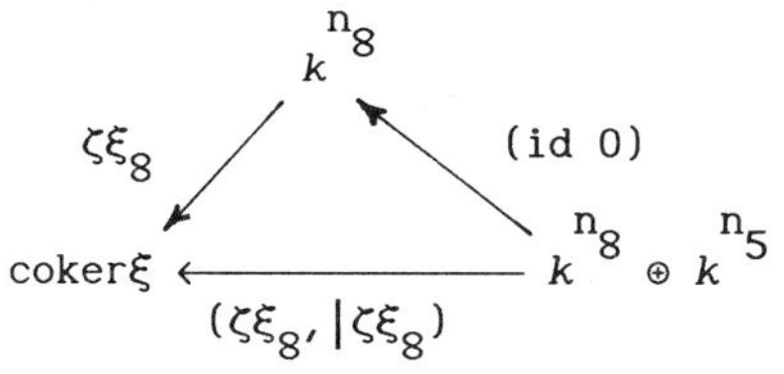

Finally, we define λ' and ρ' by setting

$$\lambda'(X,Y,Z) = (Y^t_{4'}|Y^t_4|Y^t_{3'}|Y^t_3|Y^t_2|Y^t_1|X^t_{0'}|X^t_0)^t \text{ modulo } \pi ,$$

$$\rho'(X,Y,Z) = (X_0|X_{0'}|X_1|X_2|X_3|X_{3'}|X_4|X_{4'}|X_5|X_6|X_7|X_{7'}|X_8|X_{8'}) \text{ modulo } \pi .$$

DEFINITION (The subsets $\mathfrak{M}$ and $\hat{\mathfrak{M}}$ of $\mathcal{A}$). We define $\mathfrak{M}$ to be the set of all $(X,Y,Z) \in \mathcal{A}$ which satisfy the following conditions:

(m_1) $\lambda(X,Y,Z) \in \operatorname{preproj} kQ_3$.

(m_2) $\rho(X,Y,Z) \in \operatorname{preinj} kQ_3$.

(m_3) $\lambda'(X,Y,Z)$ does not contain a direct summand which is isomorphic to one of the objects $(k,K_1,(1))$, $(k,K_2,(1))$, $(k^2,K_3,\left(\frac{1\ 0}{0\ 1}\right))$.

(m_4) $\rho'(X,Y,Z)$ does not contain a direct summand which is isomorphic to one of the objects $(K_5,k,(1))$, $(K_6,k,(1))$, $(K_7,k^2,\left(\begin{array}{c|c}1&0\\0&1\end{array}\right))$.

Moreover, we define $\hat{\mathfrak{M}}$ to be the set of all $(X,Y,Z) \in \mathcal{A}$ which satisfy the conditions (m_1), ($\hat{m}_2$), (m_3), (m_4), where ($\hat{m}_2$) is the following weakened version of (m_2):

($\hat{m}_2$) $\rho(X,Y,Z) \in \text{preinj}kQ_3 \vee (\bigvee_{\lambda\in\check{I}} \text{reg}_\lambda kQ_3)$.

We proceed by introducing mappings $\sigma : \mathcal{A} \longrightarrow F(K_1)$, $\zeta : \mathcal{A} \longrightarrow F(K)$ and $\vartheta : \mathcal{A} \longrightarrow U(\tilde{K})$, which will serve for the definition of full subcategories of $F(K_1)$, $F(K)$ and $U(\tilde{K})$. Let $(X,Y,Z) \in \mathcal{A}$ and let $\{m_j, n_i \mid i \in I_X, j \in I_Y\}$ be its set of row numbers and column numbers.

We define $\sigma : \mathcal{A} \longrightarrow F(K_1)$ by setting $\sigma(X,Y,Z)=(\bigoplus_{i=0}^{2} K_i^{\nu_i}, R_3^{\nu}, \mu_\sigma(X,Y,Z))$, where

$$\nu_0 = m_1 + m_4 + n_2 + n_4 + 2n_7 + n_8,$$
$$\nu_1 = n_0 + n_1 + n_5 + n_8,$$
$$\nu_2 = m_3 + m_4 + n_3 + n_4 + n_6 + n_7 + n_8,$$
$$\nu = m_0 + m_1 + m_2 + 2m_3 + 3m_4 + n_1 + n_4 + n_6 + 2n_7 + 2n_8,$$

and where $\mu_0(X,Y,Z)$ is given by the R_3-matrix on the following page.

We define $\zeta : \mathcal{A} \longrightarrow F(K)$ by setting $\zeta(X,Y,Z) = (\bigoplus_{i=-3}^{8} K_i^{\nu_i}, k^{\nu}, \mu_\zeta(X,Y,Z))$, where

$$\nu_i = m_{i+4} + m_4 \quad , \text{ for all } i = -3,-2,-1,$$
$$\nu_i = n_i \quad , \text{ for all } i = 0,\dots,8,$$
$$\nu = m_0 + m_1 + m_2 + 2m_3 + 3m_4,$$

and where $\mu_\zeta(X,Y,Z)$ is given by the k-matrix on page 19.

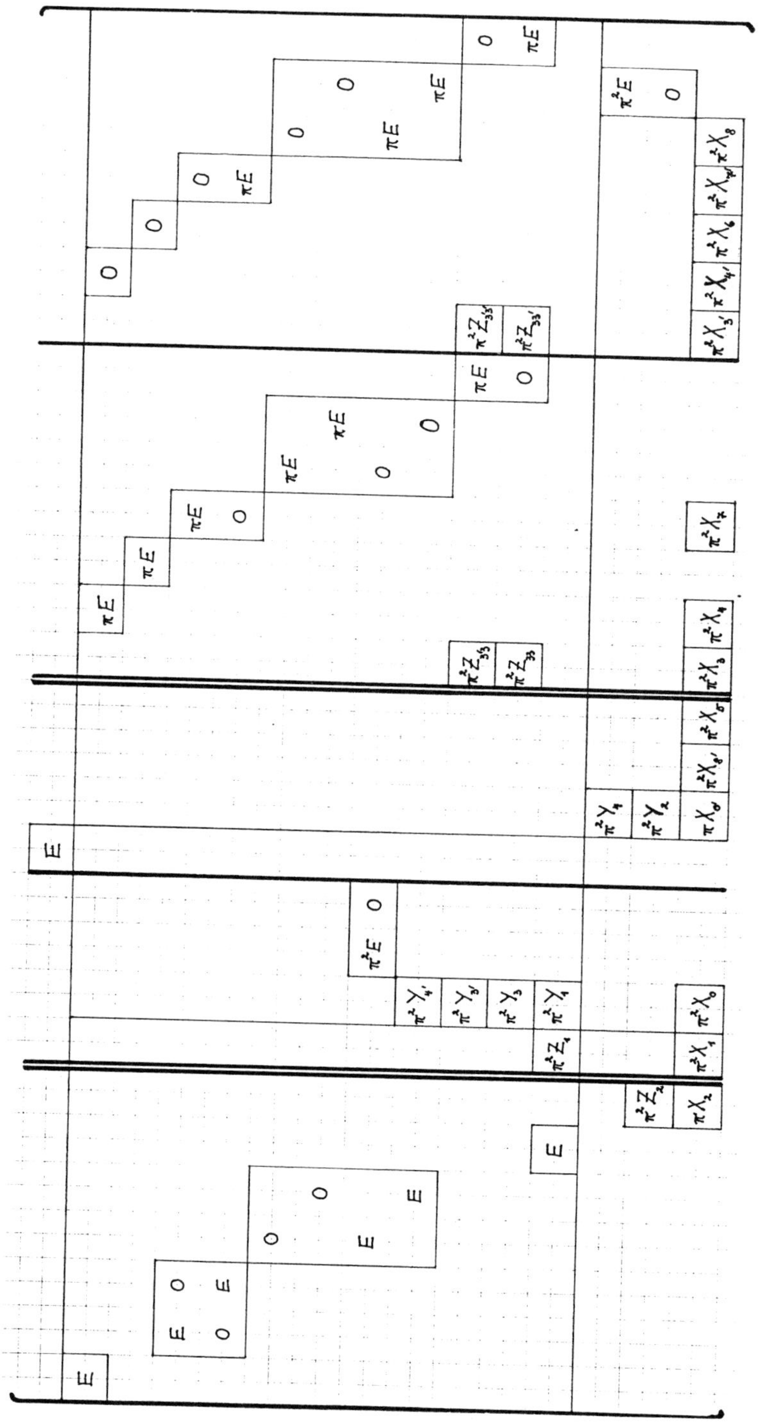

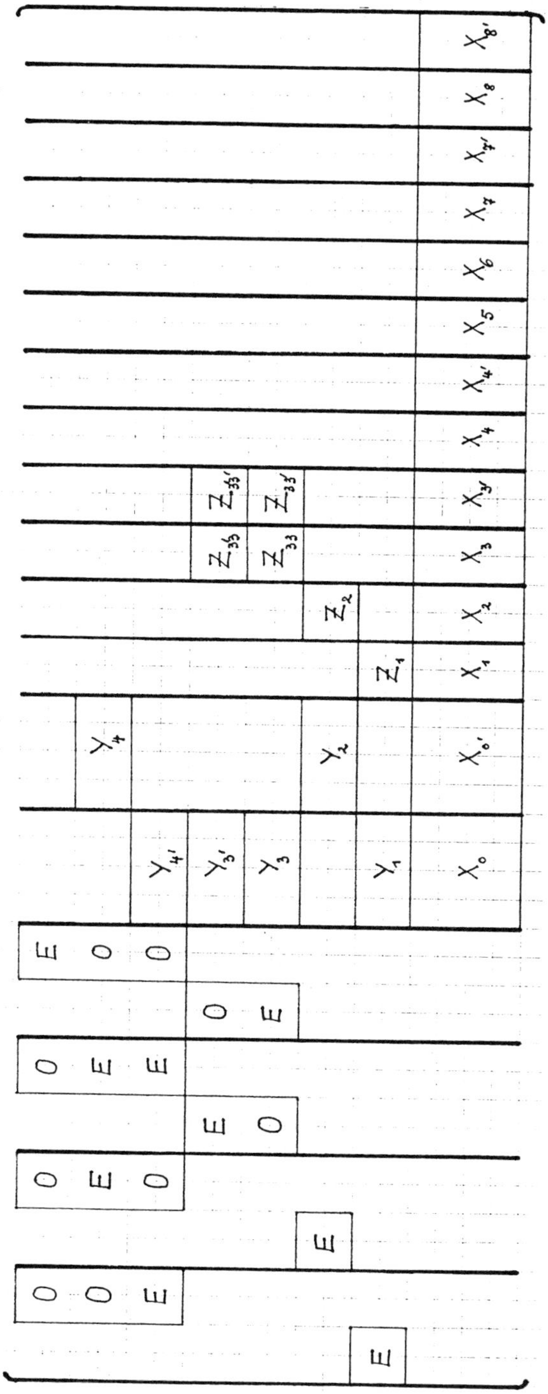

						X_8'
						X_8
						X_7'
						X_7
						X_6
						X_5
						X_4'
						X_4
		$Z_{35'}$	$Z_{33'}$			X_3'
		Z_{35}	Z_{33}			X_3
				Z_2		X_2
					Z_1	X_1
	Y_4				Y_2	X_0'
		Y_4'	Y_5' Y_5	Y_3	Y_1	X_0
E O O						
		O E				
O E E						
		E O				
O E O						
			E			
O O E						
			E			

(In $\mu_\zeta(X,Y,Z)$, the matrices X_i, Y_j, Z_k clearly have to be understood modulo π .)

We define $\vartheta : \mathcal{A} \longrightarrow U(\tilde{K})$ by setting $\vartheta(X,Y,Z) = (k^\nu, \bigoplus_{i=-3}^{8} K_i^{\nu_i}, \mu_\vartheta(X,Y,Z))$, where ν and ν_i $(i = -3,\ldots,8)$ are given as in case of ζ, and where $\mu_\vartheta(X,Y,Z) = (\mu_\zeta(X,Y,Z))^t$.

DEFINITION. (Full subcategories of $F(K_1)$, $F(K)$ and $U(\tilde{K})$) . We define the full subcategories S and $\hat{S}$ of $F(K_1)$, $\hat{C}$ of $F(K)$, $\hat{D}$ of $U(\tilde{K})$ by their classes of objects $\mathrm{Ob}S$, $\mathrm{Ob}\hat{S}$, $\mathrm{Ob}\hat{C}$, $\mathrm{Ob}\hat{D}$ in the following way:

(i) $\mathrm{Ob}S = \{M \in F(K_1) \mid M = \sigma(X,Y,Z)\ , \text{ for some } (X,Y,Z) \in \mathfrak{M}\}$

(ii) $\mathrm{Ob}\hat{S} = \{M \in F(K_1) \mid M = \sigma(X,Y,Z)\ , \text{ for some } (X,Y,Z) \in \hat{\mathfrak{M}}\}$

(iii) $\mathrm{Ob}\hat{C} = \{C \in F(K) \mid C = \zeta(X,Y,Z)\ , \text{ for some } (X,Y,Z) \in \hat{\mathfrak{M}}\}$

(iv) $\mathrm{Ob}\hat{D} = \{D \in U(\tilde{K}) \mid D = \vartheta(X,Y,Z)\ , \text{ for some } (X,Y,Z) \in \hat{\mathfrak{M}}\}$.

The connection between subsets of the set of angular matrices and the full subcategories defined above is clarified by the following commutative diagram which we include for the convenience of the reader.

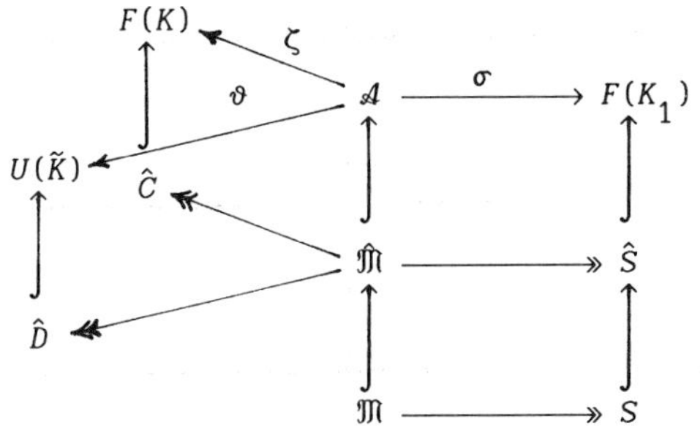

There are some additional full subcategories of $F(K_1)$ which we still have to introduce. Their definition is no longer based on angular matrices, but instead on normal forms of objects in $\mathrm{reg}_\lambda kQ_3$, as given in subsection 0.3. We denote by $\begin{array}{|c|c|}\hline R(\lambda)' & \\ \hline R(\lambda) & R(\lambda)'' \\ \hline\end{array}$ arbitrary triples of R_3-matrices which modulo π correspond to objects in $\mathrm{reg}_\lambda kQ_3$. We denote by  arbitrary R_3-matrices with entries in π^2R_3.

DEFINITION. (Full subcategories of $F(K_1)$). (i) For all $\lambda \in I \setminus \{0,\delta\}$, we define $T(\lambda)$ to be the full subcategory of $F(K_1)$ consisting of all objects (K,F,φ), for which φ is given by an R_3-matrix of the form

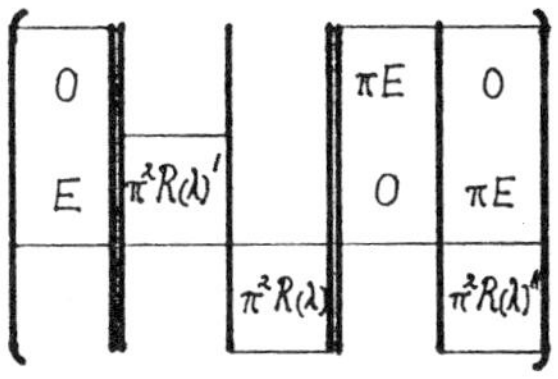

(ii) We define $T(0)$ to be the full subcategory of $F(K_1)$ consisting of all objects (K,F,φ), for which φ is given by an R_3-matrix of the form

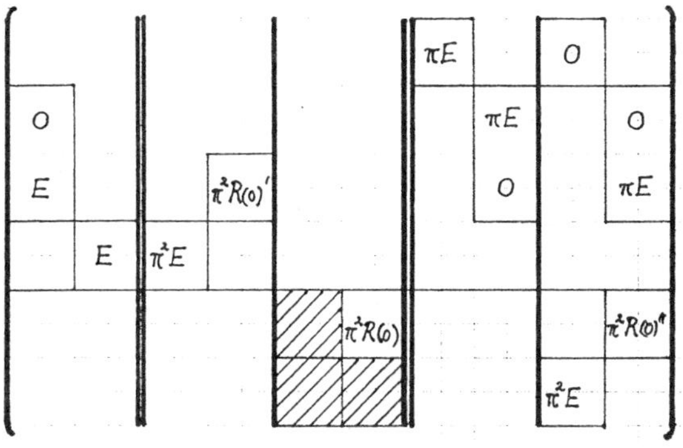

(iii) We define $T(\delta)$ to be the full subcategory of $F(K_1)$ consisting of all objects (K,F,φ), for which φ is given by an R_3-matrix of the form

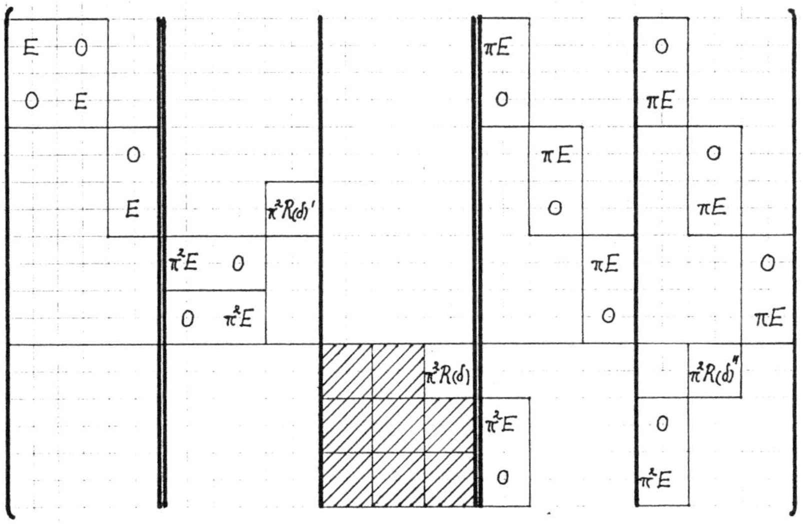

(iv) We set $T = \bigvee_{\lambda\in I} T(\lambda)$.

(v) We set $T(E) = \bigvee_{\lambda\in E} T(\lambda)$.

0.5. DIMENSION MAPPINGS.

We shall have to deal with Krull-Schmidt categories of four different types, namely with lattice categories, module categories, generalized factorspace categories and subspace categories. With each of these categories we associate a distinguished <u>dimension mapping</u> in the following way.

1) Let $_{\Xi}L$ be the lattice category over an R-order Ξ . Then $\underline{\dim}_{\Xi} : {_{\Xi}L} \longrightarrow \mathbb{N}_0$ is defined by $\underline{\dim}_{\Xi}(M) = \text{rank}_R M$, for all $M \in {_{\Xi}L}$.

2) Let modA be the module category over a k-algebra A , and let $\{S_i \mid i \in I\}$ be a set of representatives for the isomorphism classes of simple A-modules. Then $\underline{\dim}_A : \text{modA} \longrightarrow \bigoplus_I \mathbb{N}_0$ is defined by

$\underline{\dim}_A(M) = (n_i)_{i\in I}$, where n_i is the multiplicity of S_i in a composition series of M , for all $M \in \mathrm{mod}A$.

3) Let $F(K)$ be the generalized factorspace category of a Krull-Schmidt subcategory K of $\mathrm{mod}R_n$, and let $\{K_i \mid i \in I\}$ be a set of representatives for the isomorphism classes of indecomposable objects of K . Then $\underline{\dim}_{F(K)} : F(K) \longrightarrow \mathbb{N}_0 \oplus (\underset{I}{\oplus} \mathbb{N}_0)$ is defined by $\underline{\dim}_{F(K)}(K,F,\varphi) = (m;(n_i)_{i\in I})$, where $F \cong R_n^m$ and $K \cong \underset{i\in I}{\oplus} K_i^{n_i}$, for all $(K,F,\varphi) \in F(K)$.

4) Let $U(K)$ be the subspace category of a vectorspace subcategory K of $\mathrm{mod}k$, and let $\{K_i \mid i \in I\}$ be a set of representatives for the isomorphism classes of indecomposable objects of K . Then $\underline{\dim}_{U(K)} : U(K) \longrightarrow \mathbb{N}_0 \oplus (\underset{I}{\oplus} \mathbb{N}_0)$ is defined by $\underline{\dim}_{U(K)}(U,K,\psi) = (m;(n_i)_{i\in I})$, where $U \cong k^m$ and $K \cong \underset{i\in I}{\oplus} K_i^{n_i}$, for all $(U,K,\psi) \in U(K)$.

Usually it will be clear to which Krull-Schmidt category the dimension mapping refers, and then we just write $\underline{\dim}$ instead of $\underline{\dim}_\Xi$, $\underline{\dim}_A$, $\underline{\dim}_{F(K)}$ or $\underline{\dim}_{U(K)}$. Note that in all cases the dimension mapping is additive on short exact sequences. Given any Krull-Schmidt category K with dimension mapping $\underline{\dim}_K$, the value $\underline{\dim}_K(K)$ at an object $K \in K$ will be called the <u>dimension type</u> of K .

1. FIRST REDUCTION

This first reduction will be performed on the base-free level. It reduces the problem of determining $A(\Lambda)$ to the problem of determining $A(F(K_1))$.

PROPOSITION 1.1. There exists a representation equivalence $\phi_0 : {}_{\Lambda}L \longrightarrow F(K_0)$ which induces an isomorphism of Auslander-Reiten quivers $A(\Lambda) \cong A(F(K_0))$.

PROOF. This is nothing but an application of the main result from [Di 85] to our situation. Let g be a generating element of the group C_3 , and set $\bar{\Lambda} = \Lambda/(g^2+g+1)$. Then $\bar{\Lambda} \cong R[X]/(X^2+X+1) \cong R[\zeta]$, ζ being a primitive third root of unity, since by hypothesis X^2+X+1 is irreducible in K[X] . Moreover, let $|\cdot| : {}_{\bar{\Lambda}}L \longrightarrow \mathrm{mod} R_4$ be the additive functor which is defined on objects by $|N| = N/(\zeta-1)N$, and on morphisms by $|\nu| = \nu/\nu|_{(\zeta-1)N}$. With the pair $({}_{\bar{\Lambda}}L, |\cdot|)$ we associate the generalized factorspace category $F(\Lambda) = F({}_{\bar{\Lambda}}L, |\cdot|)$. By [Di 85], Theorem 1.2, there exists a representation equivalence $\phi_0 : {}_{\Lambda}L \longrightarrow F(\Lambda)$ which induces an isomorphism of Auslander-Reiten quivers $A(\Lambda) \cong A(F(\Lambda))$. Therefore it suffices to show that $F(\Lambda)$ and $F(K_0)$ are equivalent categories.

In fact, a complete set of representatives for the isomorphism classes of indecomposable $\bar{\Lambda}$-lattices is given by $\{N_i = \pi^i\bar{\Lambda} + (\zeta-1)\bar{\Lambda} \mid i = 0,1,2\}$. (This can be deduced either from the theory of Bass-orders, or by calculating the Auslander-Reiten quiver of $_{\bar{\Lambda}}L$.)

From this classification of the indecomposable $\bar{\Lambda}$-lattices it is easily seen that $|_{\bar{\Lambda}}L|$ and K_0 are equivalent subcategories of modR_4 . Hence $F(\Lambda)$ and $F(K_0)$ are equivalent categories. q.e.d.

DEFINITION. We define $\check{F}_0$ to be the full subcategory of $F(K_0)$ which consists of all objects having no direct summand isomorphic to $(\mathrm{K}_0, \mathrm{R}_4, (1))$.

PROPOSITION 1.2. There exists a representation equivalence $\phi_1 : \check{F}_0 \longrightarrow F(K_1)$ which induces an isomorphism of quivers $A_{F(K_0)}(\check{F}_0) \cong$ $\cong A(F(K_1))$.

PROOF. We proceed in two steps, proving the quiver-isomorphisms $A_{F(K_0)}(\check{F}_0) \cong A(\check{F}_0)$ and $A(\check{F}_0) \cong A(F(K_1))$.

STEP 1. The inclusion $\check{F}_0 \longrightarrow F(K_0)$ induces a quiver-monomorphism $A_{F(K_0)}(\check{F}_0) \longrightarrow A(\check{F}_0)$ which is an isomorphism if and only if $\mathrm{rad}^2_{\check{F}_0}(X,X') = \mathrm{rad}^2_{F(K_0)}(X,X')$, for all $X,X' \in \mathrm{ind}\check{F}_0$. Hence it suffices to show that for any commutative triangle

$$(\Delta) \qquad \begin{array}{ccccc} & & (1) & & \\ & {\scriptstyle\xi}\nearrow & & \searrow{\scriptstyle\xi'} & \\ X & & \xrightarrow{\quad\psi\quad} & & X' \end{array}$$

in $F(K_0)$, with $X,X' \in \mathrm{ind}\check{F}_0$ and $(1) = (\mathrm{K}_0, \mathrm{R}_4, (1))$, there exists a non-trivial factorization of ψ in $\check{F}_0$.

For this purpose consider the representation equivalence $\phi_0 : {}_{\Lambda}L \longrightarrow F(K_0)$ from Proposition 1.1. Choose a commutative triangle

(Δ′)

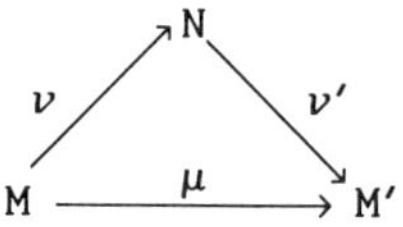

in ${}_{\Lambda}L$ which is preimage of (Δ) under ϕ_0 . Then $N \cong \Lambda$ and (Δ′) extends to a commutative diagram

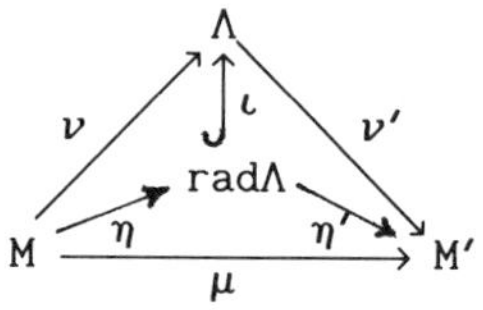

Observe that η' is a non-isomorphism. If η is also a non-isomorphism then $\psi = \phi_0(\eta')\phi_0(\eta) \in \mathrm{rad}^2_{\check{F}_0}(X,X')$. If η is an isomorphism and $\nu' = \nu_2'\nu_1' \in \mathrm{rad}^2_{\Lambda}(\Lambda,M')$ then $\psi = \phi_0(\nu_2')\phi_0(\nu_1'\iota\eta) \in \mathrm{rad}^2_{\check{F}_0}(X,X')$. If η is an isomorphism and $\nu' \notin \mathrm{rad}^2_{\Lambda}(\Lambda,M')$ then the Auslander-Reiten sequence in ${}_{\Lambda}L$ starting in radΛ is of the form $0 \longrightarrow \mathrm{rad}\Lambda \xrightarrow{\binom{\iota}{\varepsilon}} \Lambda \oplus E \xrightarrow{(\nu'\varepsilon')} M' \longrightarrow 0$, and $\psi = -\phi_0(\varepsilon')\phi_0(\varepsilon\eta) \in \mathrm{rad}^2_{\check{F}_0}(X,X')$. This proves the first isomorphism $A_{F(K_0)}(\check{F}_0) \cong A(\check{F}_0)$.

STEP 2. We introduce a functor $\phi_1 : \check{F}_0 \longrightarrow F(K_1)$. Let $X = (K,F,\varphi) \in \check{F}_0$. Then $\mathrm{im}\varphi \subset \pi F$ because otherwise, using $K_0(K_0,K) = \mathrm{Hom}_{R_4}(K_0,K)$, we conclude that $(K_0,R_4,(1))$ is a direct summand of X , contradicting $X \in \check{F}_0$. Hence φ factors through the canonical epimorphism $K \twoheadrightarrow \bar{K} = K/\pi^3K$, inducing the R_3-linear map $\tilde{\varphi} : \bar{K} \longrightarrow \pi F$. Set $\bar{F} = F/\pi^3F$ and let $\rho : \pi F \xrightarrow{\sim} \bar{F}$ be the canonical isomorphism of R_3-modules. Then we define the functor ϕ_1 on objects by $\phi_1(K,F,\varphi) = (\bar{K},\bar{F},\rho\tilde{\varphi})$, and on morphisms by $\phi_1(\tau,\sigma) = (\bar{\tau},\bar{\sigma})$, where $\bar{\tau} = \tau/\tau|_{\pi^3K}$ and $\bar{\sigma} = \sigma/\sigma|_{\pi^3F}$.

If $X = (K,F,\varphi)$, $X' = (K',F',\varphi') \in \check{F}_0$ and $\tau \in K_0(K,K')$, $\sigma \in \mathrm{Hom}_{R_4}(F,F')$, then it is easily verified that $\sigma\varphi = \varphi'\tau$ if and only if $\bar{\sigma}\rho\tilde{\varphi} = \rho'\tilde{\varphi}'\bar{\tau}$. Consequently, ϕ_1 is well-defined and full. Moreover, it is easily seen that ϕ_1 is dense and isomorphism-reflecting, hence a representation equivalence.

Therefore ϕ_1 induces a quiver-monomorphism $A(F(K_1)) \longrightarrow A(\check{F}_0)$, and this is an isomorphism if and only if $\ker(\phi_1)_{X,X'} \subset \mathrm{rad}^2_{\check{F}_0}(X,X')$, for all $X,X' \in \mathrm{ind}\check{F}_0$. Indeed, this latter condition is satisfied. Namely, if $X = (K,F,\varphi)$, $X' = (K',F',\varphi') \in \mathrm{ind}\check{F}_0$ and $\psi = (\tau,\sigma) \in \ker(\phi_1)_{X,X'}$, then $\mathrm{im}\varphi \subset \pi F$ and $\mathrm{im}\sigma \subset \pi^3F'$, hence $\sigma\varphi = 0 = \varphi'\tau$. Therefore ψ admits the following factorization.

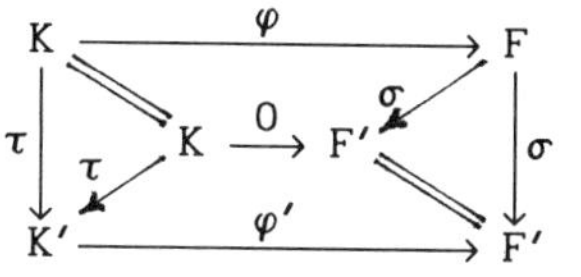

If neither X nor X' is isomorphic to any of the indecomposable objects in $I = \{(0,R,0),(K_i,0,0) \mid i = 0,1,2\}$ then $(1_K,\sigma)$ is not a splittable monomorphism and $(\tau,1_{F'})$ is not a splittable epimorphism, hence $\psi \in \mathrm{rad}^2_{\check{F}_0}(X,X')$. If either X or X' is isomorphic to an indecomposable object in I then it is straightforward to find nontrivial factorizations of ψ , proving that $\psi \in \mathrm{rad}^2_{\check{F}_0}(X,X')$ in any case. This establishes the second isomorphism $A(\check{F}_0) \cong A(F(K_1))$. q.e.d.

COROLLARY 1.3 (First reduction). The representation equivalences ϕ_0 and ϕ_1 induce an isomorphism of quivers $A_s(\Lambda) \cong A(F(K_1))$.

PROOF. Because $\phi_0(\Lambda) = (K_0,R_4,(1))$, Proposition 1.1 implies that ϕ_0 induces an isomorphism of quivers $A_s(\Lambda) \cong A_{F(K_0)}(\check{F}_0)$. On the other hand, $A_{F(K_0)}(\check{F}_0) \cong A(F(K_1))$, by Proposition 1.2. q.e.d.

2. SECOND REDUCTION

In this second reduction we exhibit a dense full subcategory $\hat{S} \vee T(E)$ of $F(K_1)$. As a consequence, the problem of determining $A(F(K_1))$ will be reduced to the problem of determining $A_{F(K_1)}(\hat{S} \vee T(E))$. Our approach will be to view $F(K_1)$ base-dependently and to reduce matrices. We begin by defining the base-dependent equivalent of $F(K_1)$, the matrix category $\mathcal{M}$.

DEFINITION (The matrix category $\mathcal{M}$). Objects in $\mathcal{M}$ are matrices $M = (A_0 \parallel \pi^2 A_1 \mid A_2 \parallel \pi A_3 \mid \pi A_4) \in R_3^{m\times n}$, where $A_i \in R_3^{m\times n_i}$, $i = 0,\dots,4$, and $\{m, n_0, \dots, n_4\} \subset \mathbb{N}_0$, subject to $n_1 = n_2$, $n_3 = n_4$, $\sum_{i=0}^{4} n_i = n$. Morphisms in $\mathcal{M}$ from M to M′ are given by pairs of matrices $(T,S) \in R_3^{n'\times n} \times R_3^{m'\times m}$ which satisfy the following conditions:

(i) $$T = \left(\begin{array}{c||c|c||c|c} T_{00} & 0 & \pi T_{02} & \pi^2 T_{03} & \pi^2 T_{04} \\ \hline\hline T_{10} & T_{11} & T_{12} & T_{13} & T_{14} \\ \hline T_{20} & \pi^2 dT_{12} & T_{11} & \pi dT_{14} & \pi T_{13} \\ \hline\hline T_{30} & \pi T_{31} & T_{32} & T_{33} & T_{34} \\ \hline T_{40} & \pi dT_{32} & T_{31} & dT_{34} & T_{33} \end{array}\right)$$

(The block-partition of T is understood to correspond to the block-partitions of M and M′).

(ii) $SM = M'T$.

Observe that $F(K_1)$ and $\mathcal{M}$ are equivalent categories. We shall be free to identify them, if it is convenient. In particular, we consider $S, T, \hat{S}, T(E)$ etc. as full subcategories of $\mathcal{M}$. (See subsection 0.4 for the definition of subcategories.)

PROPOSITION 2.1. The full subcategory $S \vee T$ of $\mathcal{M}$ is dense in $\mathcal{M}$.

PROOF. Let $M^{(0)}$ be an arbitrary object in $\mathcal{M}$. We have to show that $M^{(0)}$ is isomorphic to an object in $S \vee T$.

For this purpose we deal with the matrix problem associated with $\mathcal{M}$. The matrices M of the matrix problem are the objects of $\mathcal{M}$. The admissible transformations on M are given by the isomorphisms of $\mathcal{M}$. We say that M and M′ are equivalent matrices (and write $M \sim M'$) if M can be transformed to M′ by means of admissible transformations. Starting with $M^{(0)}$, we shall construct a sequence of equivalences $M^{(0)} \sim M^{(1)} \sim \dots \sim M^{(6)}$, such that $M^{(6)} \in S \vee T$.

Concerning this sequence of equivalences some general remarks are necessary. By the "n-th step" $(n = 1,\dots,6)$ we mean the passage from $M^{(n-1)}$ to $M^{(n)}$. This will always be achieved by operating on $M^{(n-1)}$ only with "induced" admissible transformations (i.e. admissible transformations on $M^{(n-1)}$ stabilizing all submatrices of $M^{(n-1)}$ which are in normal form). After the n-th step we hatch the part $H^{(n)}$ of $M^{(n)}$ which is not in normal form so far, according to the following convention:

submatrices hatched like 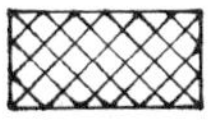 have entries in R_3 ,

submatrices hatched like 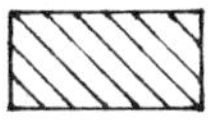 have entries in πR_3 ,

submatrices hatched like 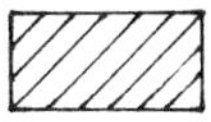 have entries in $\pi^2 R_3$.

Submatrices which are neither hatched nor of an indicated normal form are understood to be zero matrices.

Moreover, after the n-th step we introduce coordinates for relevant subsets of the set of rows of $M^{(n)}$, columns of $M^{(n)}$ respectively, and we agree to use them in the following way: $M^{(n)}(i,j)$ denotes the submatrix of $M^{(n)}$ which is given by the intersection of row strip i and column strip j. It is understood that the union of row strips i and i' (column strips j and j' respectively) has the coordinate $\frac{i}{i'}$ $(j|j'$ respectively). Accordingly we put

$$M^{(n)}\left(\frac{\frac{i_1}{\vdots}}{i_\ell}, j_1|\dots|j_m\right) = \begin{pmatrix} M^{(n)}(i_1,j_1) & \dots & M^{(n)}(i_1,j_m) \\ \vdots & & \vdots \\ M^{(n)}(i_\ell,j_1) & \dots & M^{(n)}(i_\ell,j_m) \end{pmatrix} .$$

Likewise it will be convenient to use coordinates for indicating specific types of induced admissible transformations, as follows: $(i,j) \longrightarrow (i',j)$ means that there are induced admissible transformations on $M^{(n)}$ which effect addition of linear combinations of rows of $M^{(n)}(i,j)$ to rows of $M^{(n)}(i',j)$. $(i,j') \longleftarrow (i,j)$ means that there are induced admissible transformations on $M^{(n)}$ which effect addition of linear combinations of colums of $M^{(n)}(i,j)$ to columns of $M^{(n)}(i,j')$. Note that quite frequently induced admissible transformations of these (and related) types will have simultaneous effect on submatrices X of $M^{(n)}$, where $X \subset H^{(n)} \setminus (M^{(n)}(i,j|j') \cup M^{(n)}(\frac{i}{i'},j))$. On the one hand, if at that stage of the proof X is evidently of no concern then we will not mention the simultaneous effect on X. On the other hand, if we have to be careful about X then the simultaneous effect on X will be noted by saying "$(i,j) \longrightarrow (i',j)$, with change of X", or "$(i,j') \longleftarrow (i,j)$, with change of X".

Let $N \in \mathbb{N}$ and let $S = (M^{(n)}(i_\nu, j_\nu)_{\nu=1,\ldots,N}$ be an N-tuple of pairwise disjoint submatrices of $H^{(n)}$, $M^{(n)}(i_\nu, j_\nu)$ having entries in $\pi^{e_\nu} R_3$ for each $\nu = 1,\ldots,N$. By the local problem of S we mean the problem of transforming S to normal form, modulo $(\pi^{e_\nu+1})_{\nu=1,\ldots,N}$, by means of induced admissible transformations of $M^{(n)}$. In the crucial places of our matrix problem, the local problems are just the classification problems of Q_1, Q_2, Q_3, and S_1, S_2 (i.e. the problems of classifying all indecomposable k-representations of the quivers Q_i or species S_j). Freely and without particular reference we shall use the solutions given in 0.3 to these classification problems, substituting S by the normal form of a general k-representation of Q_i or of S_j, modulo $(\pi^{e_\nu+1})_{\nu=1,\ldots,N}$. (Note that in this procedure we actually have to pass from k-matrices to R_3-matrices. Thus, if $(A_\nu)_{\nu=1,\ldots,N}$ is the normal form of a k-representation of Q_i or of S_j then, to be precise, we substitute S by an R_3-matrix $(\pi^{e_\nu}\tilde{A}_\nu + \pi^{e_\nu+1} X_\nu)_{\nu=1,\ldots,N}$, where $(\tilde{A}_\nu)_{\nu=1,\ldots,N}$ modulo π equals $(A_\nu)_{\nu=1,\ldots,N}$ and $(X_\nu)_{\nu=1,\ldots,N}$ is an undetermined N-tuple of R_3-matrices. However, for simplicity of notation, we always put $(\pi^{e_\nu} A_\nu)_{\nu=1,\ldots,N} := (\pi^{e_\nu}\tilde{A}_\nu + \pi^{e_\nu+1} X_\nu)_{\nu=1,\ldots,N}$.)

By the reduction of $S = (M^{(n)}(i_\nu, j_\nu))_{\nu=1,\ldots,N}$ we mean the solution of the local problem of S (and the replacement of S by the normal form corresponding to this solution), followed by the transformation of parts of $H^{(n)}$ to zero by using induced admissible transformations of type $(i_\nu, j') \leftarrow (i_\nu, j_\nu) \rightarrow (i', j_\nu)$, $\nu=1,\ldots,N$. Dually, by the coreduction of S we mean the transformation of parts of S to zero by using induced admissible transformations of type

$(i',j_\nu) \to (i_\nu,j_\nu) \leftarrow (i_\nu,j')$, $\nu=1,\dots,N$, followed by the solution of the local problem of S (and the replacement of S by the normal form corresponding to this solution). Here, i' and j' are understood to denote such row strips and column strips of $M^{(n)}$ that the submatrices $M^{(n)}(i',j_\nu)$ and $M^{(n)}(i_\nu,j')$ are contained in $H^{(n)}\backslash S$.

After these comments let us now construct explicitly the chain of equivalences $M^{(0)} \sim M^{(1)} \sim \dots \sim M^{(6)}$. For each $n = 1,\dots,6$, the n-th step will consist in the reduction or coreduction of an N-tuple of submatrices of $M^{(n-1)}$. This will occupy the remaining part of the present section.

Let $M^{(0)}$ be an arbitrary matrix in $\mathcal{M}$.

$M^{(0)}$ =

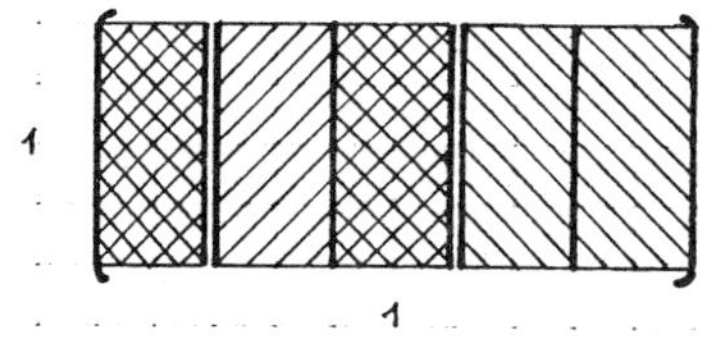

STEP 1. Reduction of $M^{(0)}(1,1)$.

The local problem of the (single) matrix $M^{(0)}(1,1)$ is the classification problem of the quiver Q_1 . Therefore $M^{(0)}$ is equivalent to the following matrix $M^{(1)}$.

$M^{(1)}$ =

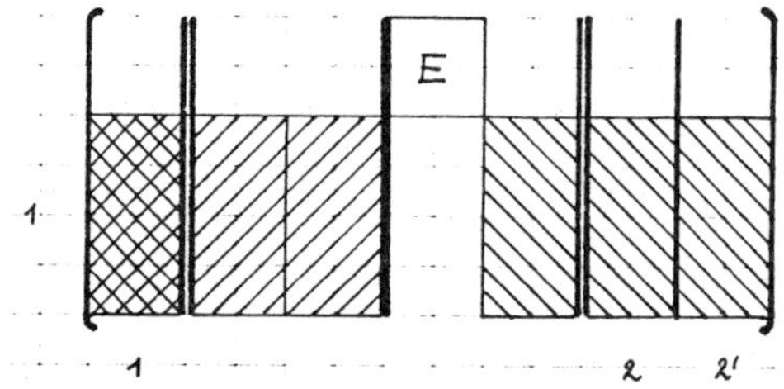

STEP 2. Reduction of $M^{(1)}(1,1|2|2')$.

The local problem of the triple of matrices $M^{(1)}(1,1|2|2')$ is the classification problem of the species S_1 . Therefore $M^{(1)}$ is equivalent to the matrix $M^{(2)}$ on the following page.

STEP 3. Reduction of $M^{(2)}(1,1|1')$.

The local problem of the pair of matrices $M^{(2)}(1,1|1')$ is the classification problem of the quiver Q_2 . We denote by $(K|K')$ the normal form of a general Kronecker-module <u>which contains no semisimple direct summand</u>. Then, $M^{(2)}$ is equivalent to the matrix $M^{(3)}$ on page 35.

STEP 4. Coreduction of $M^{(3)}(1,7)$ and $M^{(3)}(4,8)$.

On submatrices of $M^{(3)}$ there are the following induced admissible transformations:

(1) $\begin{pmatrix} \frac{3'}{\frac{3}{2}}, 8 \end{pmatrix} \longrightarrow (4,8)$, with change of $(4,\kappa)$ and $(1,\kappa'|8')$.

(2) $\begin{matrix}(\kappa,\kappa) \\ (\kappa,\pi\kappa')\end{matrix} \rightrightarrows (4,\kappa)$, with change of $(1,\kappa')$.

(3) If $M^{(3)}(4,\kappa) = 0$ then $(1,7) \longleftarrow (1,1|2|3|3'|4|4'|5|6|6')$, with change of $(\kappa,7)$ and $(1,8')$.

(4) $(\kappa,7) \longleftarrow (\kappa,\kappa|\pi\kappa')$.

Using (1) - (4) (in the indicated order), we obtain that $M^{(3)}$ is equivalent to the matrix $M^{(4)}$ on page 36.

In the next step we shall focus upon the following three distinguished submatrices of $M^{(4)}$:

the <u>vertical</u> submatrix $M_v^{(4)} := M^{(4)}\begin{pmatrix} \frac{3'}{\frac{3}{2}}, v \end{pmatrix}$,

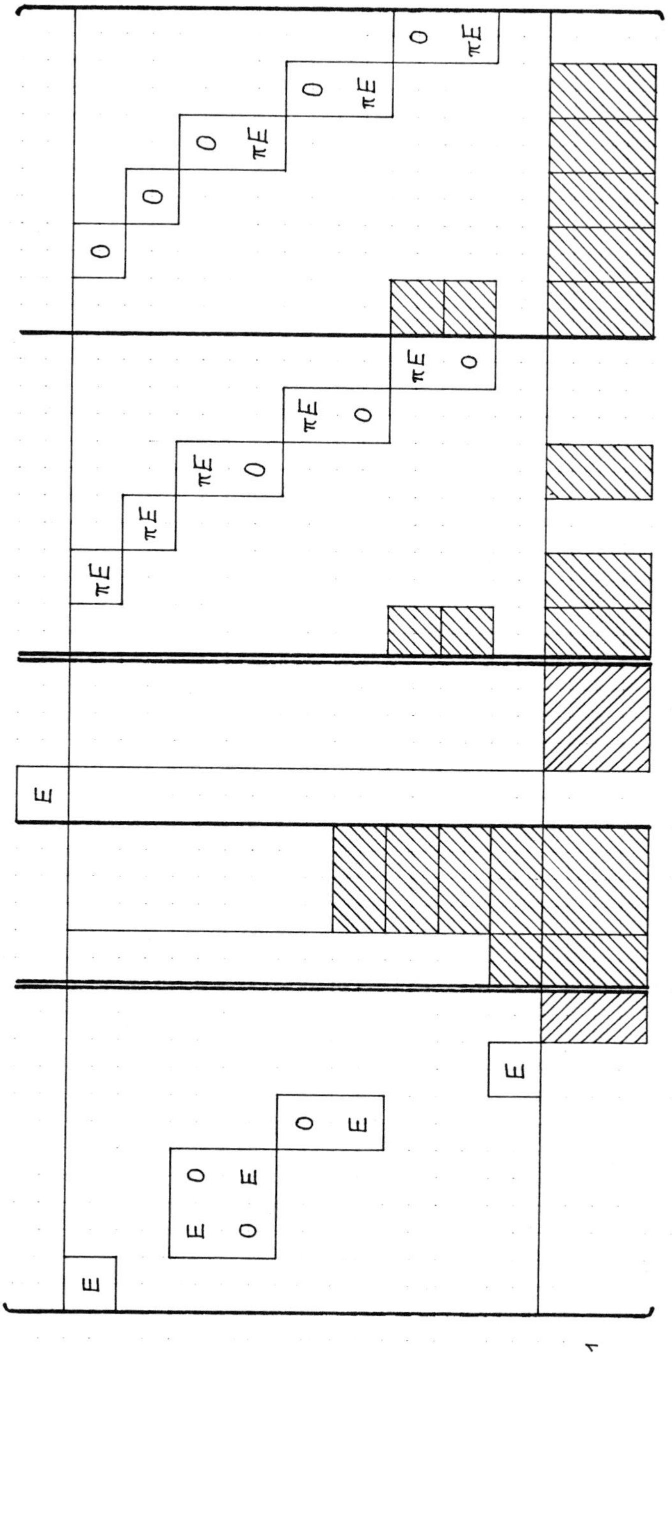
E
O
πE
1
1'

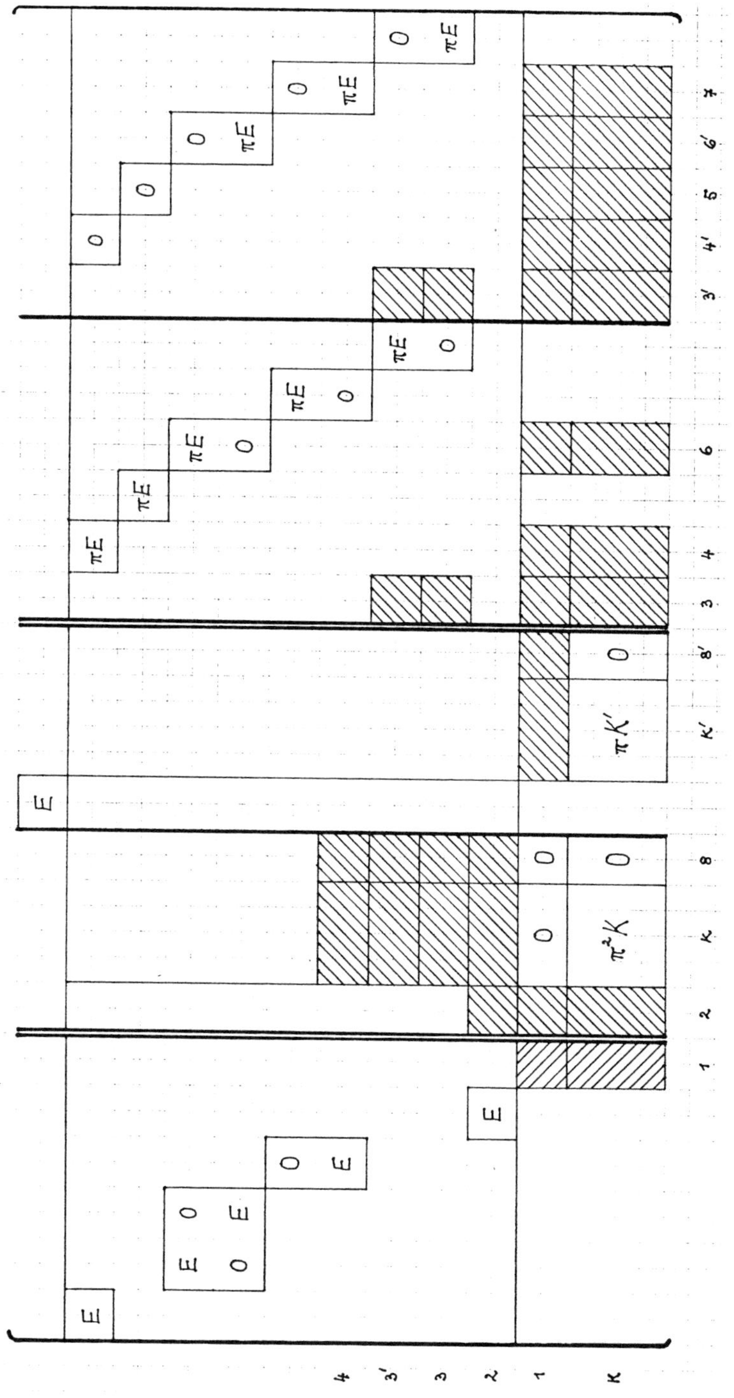
E
0
πE
π²K
πK'
1
2
K
8
K'
8'
3
4
6
3'
4'
5
6'
7
4
3'
3
2
1
K

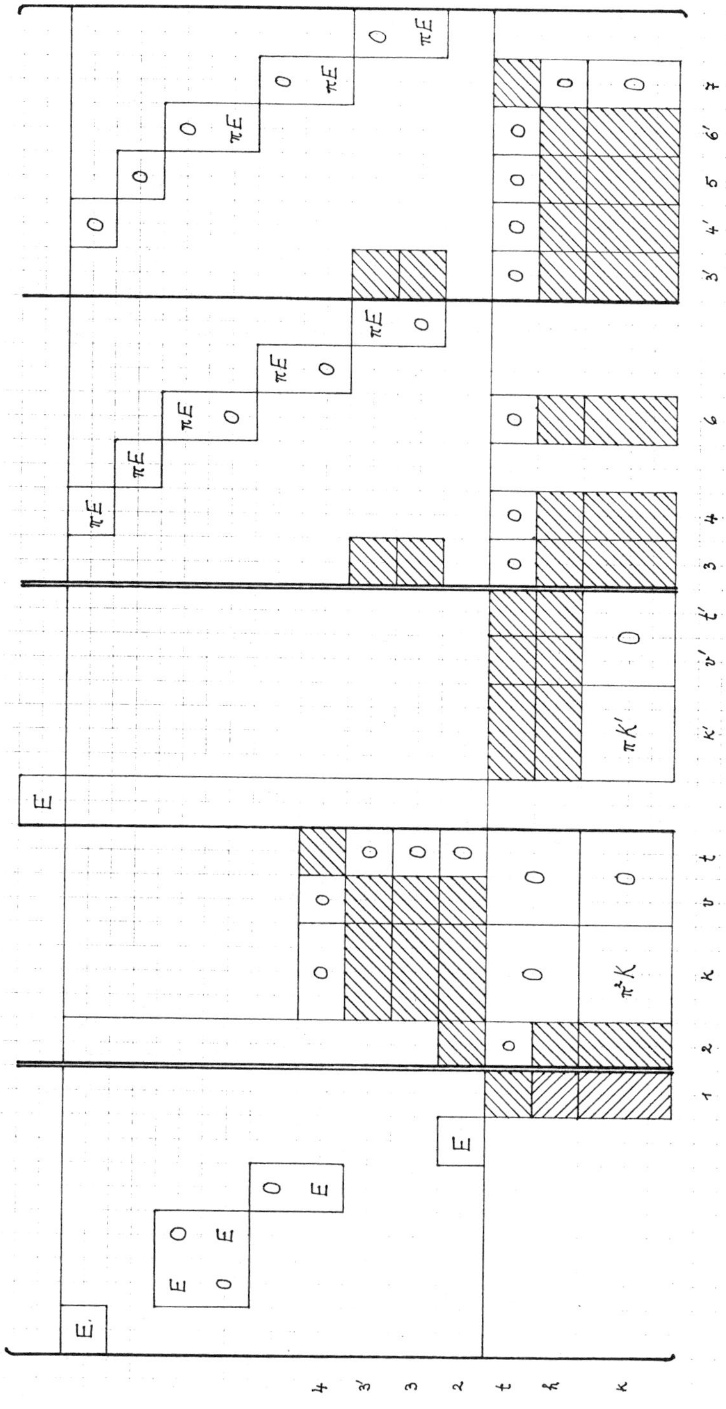

the <u>horizontal</u> submatrix $M_h^{(4)} := M^{(4)}(h\ ,\ 1|2|3|3'|4|4'|5|6|6')$,

the <u>triangular</u> submatrix $M_t^{(4)} := \begin{array}{|c|c}\hline M^{(4)}(4,t) & \\ \hline M^{(4)}(t,t') & \multicolumn{1}{|c|}{M^{(4)}(t,7)} \\ \hline \end{array}$

Note that we may assume without loss of generality that the k-linear map corresponding to $M_v^{(4)}$ is injective and the k-linear map corresponding to $M_h^{(4)}$ is surjective.

STEP 5. Reduction of $M_v^{(4)}$, $M_h^{(4)}$ and $M_t^{(4)}$.

5.1. The local problem of the triple of matrices $M_v^{(4)}$ is the classification problem of the species S_2 . Therefore $M_v^{(4)}$ is equivalent to the following matrix $M_v^{(5)}$.

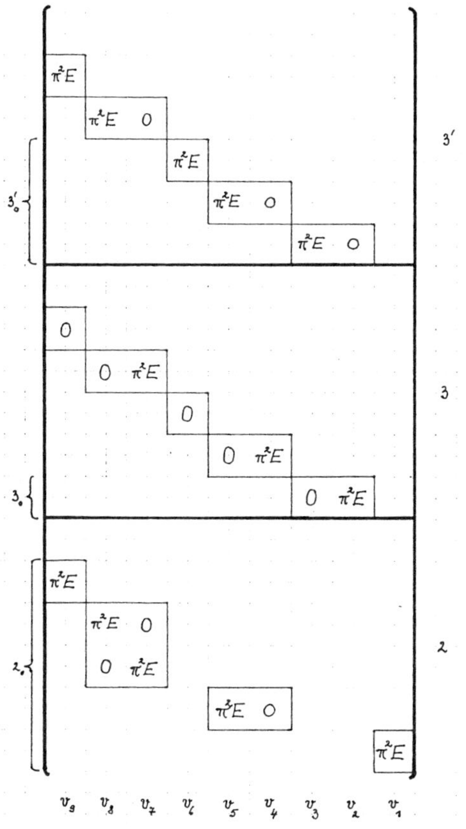

5.2. The local problem of the triple of matrices $M^{(4)}(h\ ,\ 5|6|6')$ is the classification problem of the species S_1 . The local problem of the pair of matrices $M^{(4)}(h\ ,\ 4|4')$ is the classification problem of the quiver Q_2 . With respect to k-representations of Q_2 we introduce the following notation:

$(P|P')$ = normal form of an object in $\text{preproj}\,kQ_2$,

$(R|R')$ = normal form of an object in $\bigvee_{\lambda\in I\setminus\{\infty\}} \text{reg}_\lambda kQ_2$,

$(R(\infty)|R(\infty)')$ = normal form of an object in $\text{reg}_\infty kQ_2$,

$(I|I')$ = normal form of an object in $\text{preinj}\,kQ_2$.

Then we can reduce $M_h^{(4)}$ to a partial normal form such that $M_h^{(4)}$ is equivalent to the matrix $M_h^{(5)}$ on the following double page.

5.3. The local problem of the triple of matrices $M_t^{(4)}$ is the classification problem of the quiver Q_3 . With respect to k-representations of Q_3 we introduce the following notation:

$\mathcal{P}'$	
$\mathcal{P}$	$\mathcal{P}''$

= normal form of an object in $\text{preproj}\,kQ_3$,

$\mathcal{R}'$	
$\mathcal{R}$	$\mathcal{R}''$

= normal form of an object in $\bigvee_{\lambda\in I\setminus\{0,\delta\}} \text{reg}_\lambda kQ_3$,

$\mathcal{R}(0)'$	
$\mathcal{R}(0)$	$\mathcal{R}(0)''$

= normal form of an object in $\text{reg}_0 kQ_3$,

$\mathcal{R}(\delta)'$	
$\mathcal{R}(\delta)$	$\mathcal{R}(\delta)''$

= normal form of an object in $\text{reg}_\delta kQ_3$,

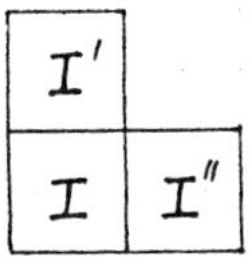

= normal form of an object in preinjkQ_3 ,

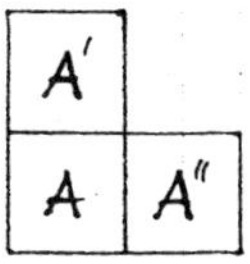

= normal form of an object in modkQ_3 .

In this terminology we have:

$M_t^{(4)} \sim$

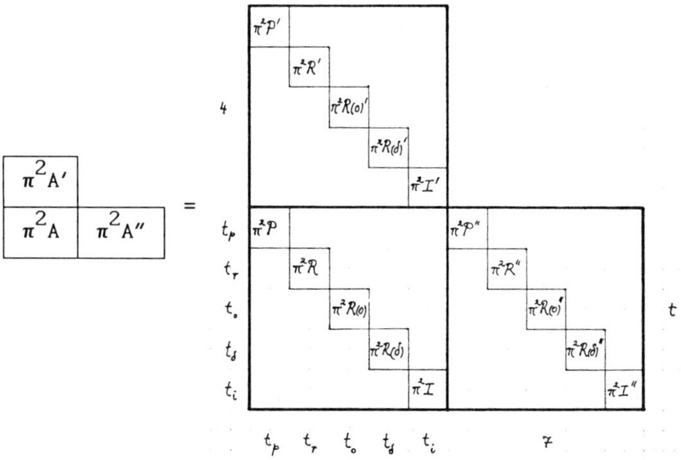

$= M_t^{(5)}$.

5.4. Define $M^{(5)}$ to be the matrix which is obtained from $M^{(4)}$ by replacing $M_v^{(4)}$, $M_h^{(4)}$ and $M_t^{(4)}$ by $M_v^{(5)}$, $M_h^{(5)}$ and $M_t^{(5)}$. Because the reductions 5.1, 5.2 and 5.3 can be carried out independently, we obtain that $M^{(4)} \sim M^{(5)}$.

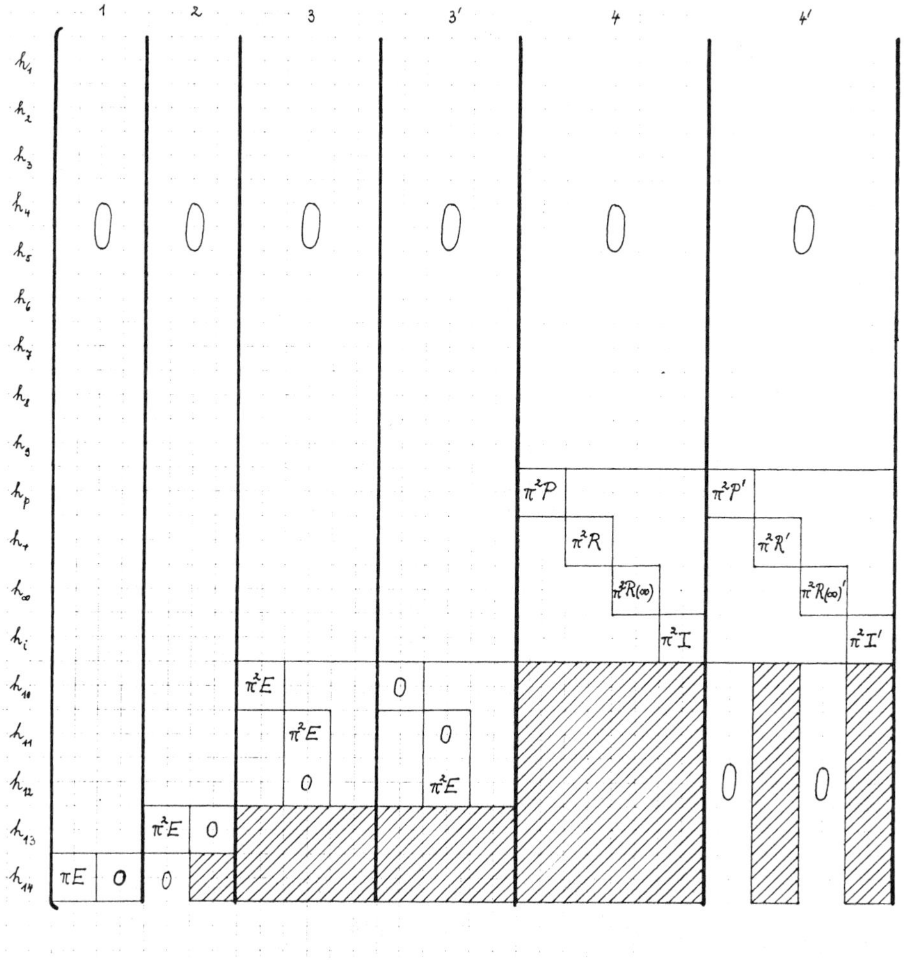
1
2
3
3'
4
4'
h1
h2
h3
h4
h5
h6
h7
h8
h9
hp
hr
h∞
hi
h10
h11
h12
h13
h14
0
π²P
π²R
π²R(∞)
π²I
π²P'
π²R'
π²R(∞)'
π²I'
π²E
πE

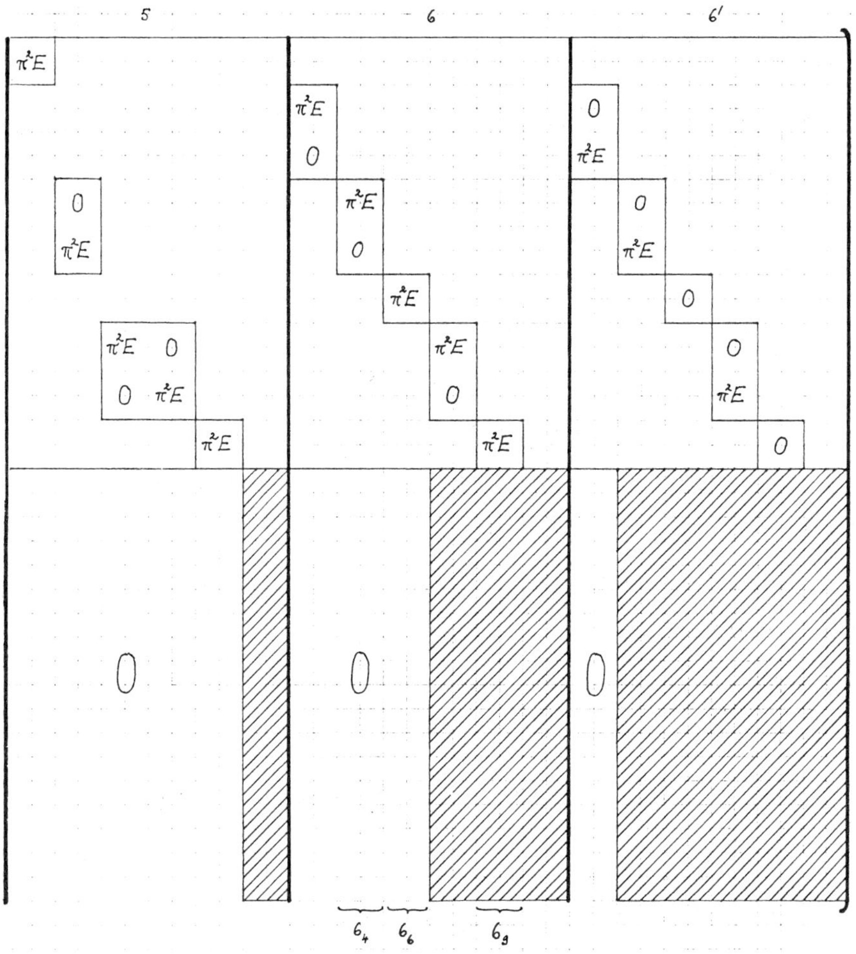
5
6
6'
π^2E
0
π^2E
π^2E 0
0 π^2E
π^2E
0
π^2E
0
π^2E
0
π^2E
π^2E
0
π^2E
0
π^2E
0
π^2E
0
0
π^2E
0
0
0
0
6_4
6_6
6_9

STEP 6. Coreduction of $M_s^{(5)}$.

In this last and most involved step we focus upon the submatrix $M_s^{(5)} := M^{(5)} (\frac{t}{h}, \kappa'|v'|t')$ of $M^{(5)}$. It has to be emphasized that the arguments (1) - (29) used in this coreduction can be established only when working with the explicit form of $M^{(5)}$. Since this matrix is too big for reproduction in a Journal, the interested reader (in case he has enough patience and a sufficiently large sheet of paper) unfortunately has to write it down by himself.

6.1. Refinement of coordinates.

We keep the coordinates of $M^{(4)}$ for $M^{(5)}$ and we refine the coordinates t, h, v', t' in accordance with the coordinates of $M_v^{(5)}$, $M_h^{(5)}$ and $M_t^{(5)}$, as given in 5.1, 5.2 and 5.3. In particular we write $v' = (v_9|\ldots|v_1)$ and $t' = (t_p|\ldots|t_i)$, omitting dashes in the refined coordinates for simplicity.

For the row coordinates t and h we introduce the following further refinements:

$t_{p_n,\omega}$ denotes the substrip of t_p which corresponds to the last row in $\pi^2 P_n$, $n \in \mathbb{N}_0$;

$t_{0_n,\alpha}$ denotes the substrip of t_0 which corresponds to the first row in $\pi^2 R(0)_n$, $n \in \mathbb{N}$;

$t_{\delta_n,\omega}$ denotes the substrip of t_δ which corresponds to the last row in $\pi^2 R(\delta)_n$, $n \in \mathbb{N}$;

$h_{p_n,\omega}$ denotes the substrip of h_p which corresponds to the last row in $\pi^2 P_n$, $n \in \mathbb{N}_0$;

$t_{p,\omega} := \bigcup_{n\in\mathbb{N}_0} t_{p_n,\omega}$; $t_{0,\alpha} := \bigcup_{n\in\mathbb{N}} t_{0_n,\alpha}$; $t_{\delta,\omega} := \bigcup_{n\in\mathbb{N}} t_{\delta_n,\omega}$;

$h_{p,\omega} := \bigcup_{n\in\mathbb{N}_0} h_{p_n,\omega}$.

For the column coordinates κ' and t' we introduce the following further refinements:

$\kappa'_{0_n,\omega}$ denotes the substrip of κ' which corresponds to the last column in $\pi R(0)'_n$, $n\in\mathbb{N}$;

$\kappa'_{i_n,\alpha}$ denotes the substrip of κ' which corresponds to the first column in $\pi I'_n$, $n\in\mathbb{N}$;

$t_{0_n,\omega}$ denotes the substrip of t_0 which corresponds to the last column in $\pi^2 R(0)_n$, $n\in\mathbb{N}$;

$t_{\delta_n,\alpha}$ denotes the substrip of t_δ which corresponds to the first column in $\pi^2 R(\delta)_n$, $n\in\mathbb{N}$;

$t_{i_n,\omega}$ denotes the substrip of t_i which corresponds to the last column in $\pi^2 I_n$, $n\in\mathbb{N}$;

$\kappa'_{0,\omega} := \bigcup_{n\in\mathbb{N}} \kappa'_{0_n,\omega}$; $\kappa'_{i,\alpha} := \bigcup_{n\in\mathbb{N}} \kappa'_{i_n,\alpha}$; $t_{0,\omega} := \bigcup_{n\in\mathbb{N}} t_{0_n,\omega}$;

$t_{\delta,n} := \bigcup_{n\in\mathbb{N}} t_{\delta_n,\alpha}$; $t_{i,\omega} := \bigcup_{n\in\mathbb{N}_0} t_{i_n,\omega}$.

6.2. Preparatory remarks.

We may assume that $M^{(5)}(t\backslash t_p,1) = 0$ and $M^{(5)}\left(\begin{array}{c} 3'_0 \\ \hline 3_0 \\ \hline 2_0 \end{array} , 2|\kappa\right) = 0$.

(See step 5.1 for definition of the row coordinates $3'_0$, 3_0 and 2_0 .)

For verification of the induced admissible transformations (1) - (29) below the following observations (i) - (iii) are useful.

(i) Arbitrary changes of $M^{(5)}(h,1)$ by multiples of π^2 can be restored in view of $(h,1) \leftarrow (h\ ,\ (\pi\cdot 1)|2|3|3'|4|4'|5|6|6')$.

(ii) Arbitrary changes of $M^{(5)}(\kappa,\kappa'|v'|t'|7)$ by multiples of π^2 can be restored in view of $(\kappa,\kappa'|v'|t'|7) \leftarrow (\kappa\ ,\ \kappa|\pi\kappa')$.

(iii) Arbitrary changes of $M^{(5)}(4,\kappa)$ by multiples of π^2 can be restored, with change of $M^{(5)}(t,\kappa')$, in view of $\begin{matrix}(\kappa,\kappa)\\(\kappa,\pi\kappa')\end{matrix} \rightrightarrows (4,\kappa)$.

6.3. Splitting off $\bigvee_{\lambda\in I\setminus\{0,\delta\}} T(\lambda)$.

There are the following induced admissible transformations on submatrices of $M^{(5)}$:

(1) $M^{(5)}\left(\begin{array}{c}h_2\\ \hline h_3\end{array},\ t'\right) \sim M^{(5)}\left(\begin{array}{c}h_2\\ \hline h_3\end{array},\ t'\right) + \pi^2\left(\begin{array}{c}C_2A - C_3A''A'\\ \hline -C_2A''A' + dC_3A\end{array}\right)$, with change of $M^{(5)}(h,\kappa'|v')$, for arbitrary coefficient matrices C_2 and C_3 .

(2) $M^{(5)}\left(\begin{array}{c}h_4\\ \hline h_5\end{array},\ t'\right) \sim M^{(5)}\left(\begin{array}{c}h_4\\ \hline h_5\end{array},\ t'\right) + \pi^2\left(\begin{array}{c}CA\\ \hline -CA''A'\end{array}\right)$, with change of $M^{(5)}(h,\kappa'|v')$ and $M^{(5)}\left(\begin{array}{c}h_p\\ \hline \vdots\\ \hline h_{14}\end{array},\ t'\right)$, for arbitrary coefficient matrix C .

(3) $(t,t') \longrightarrow (h_6,t')$, with change of $M^{(5)}(h,\kappa'|v')$ and $M^{(5)}\left(\begin{array}{c}h_p\\ \hline \vdots\\ \hline h_{14}\end{array},\ t'\right)$.

(4) $(h,\kappa'|v'|t') \leftarrow (h\ ,\ (\pi\cdot 1)|3|3'|4)$.

(5) $(t,t') \longrightarrow \left(\begin{array}{c}h_1\\ \hline h_5\\ \hline h_7\\ \hline h_8\\ \hline h_9\\ \hline h_{p,\omega}\\ \hline h_\infty\\ \hline h_{13}\end{array},\ t'\right)$, with change of $M^{(5)}(h,\kappa'|v')$.

(6) $M^{(5)}(t,v_3|v_2) \sim M^{(5)}(t,v_3|v_2) + \pi^2(A''A'D_2 + AdD_3|AD_2 + A''A'D_3)$, with change of $M^{(5)}(h,v')$, for arbitrary coefficient matrices D_2 and D_3.

(7) $M^{(5)}(t,v_5|v_4) \sim M^{(5)}(t,v_5|v_4) + \pi^2(A''A'D|AD)$, with change of $M^{(5)}(h,v')$ and $M^{(5)}(t,\kappa')$, for arbitrary coefficient matrix D.

(8) $(t,v_6) \leftarrow (t,t')$, with change of $M^{(5)}(h,v')$ and $M^{(5)}(t,\kappa')$.

(9) $(t,v_9|v_8|v_7|v_5|v_1) \leftarrow (t,t')$, with change of $M^{(5)}(h,v')$.

(10) $(\kappa,\pi\kappa') \longrightarrow (\frac{t}{h}, \kappa')$.

(11) $(t,\kappa'_{0_1,\omega}) \leftarrow (t,\kappa'_{0_2,\omega}) \leftarrow \ldots \leftarrow (t,\kappa'_{i_2,\alpha}) \leftarrow (t,\kappa'_{i_1,\alpha}) \leftarrow (t,t')$, with change of $M^{(5)}(h,\kappa')$.

Applying (1) - (11) to $M_s^{(5)}$ we obtain that

$M_s^{(5)} \sim$

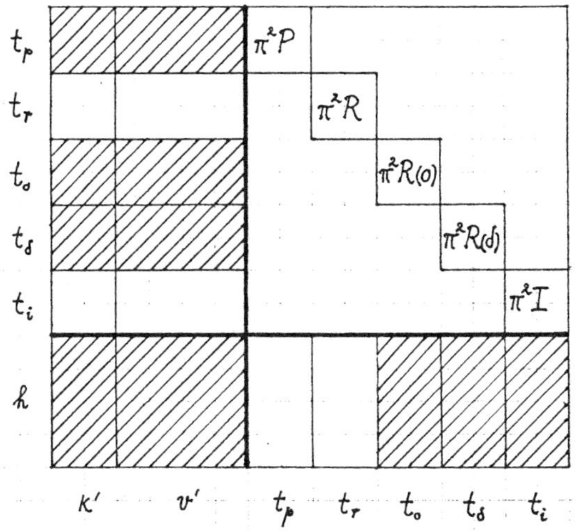

$= M_s^{(5)'}$.

Now it is evident that $\pi^2 R$ splits off from $M_s^{(5)}$. Therefore the submatrix of $M^{(5)}$ which corresponds to $\pi^2 R$ splits off from $M^{(5)}$, and this submatrix is indeed an object in $\bigvee_{\lambda\in I\setminus\{0,\delta\}} T(\lambda)$.

6.4. Splitting off $T(0)$.

We carry on with the coordinates of $M_s^{(5)\prime}$. Due to (1) - (11) we can assume that $M^{(5)}(t_p\backslash t_{p,\omega}\ ,\ v_1) = 0$, $M^{(5)}(t_0\backslash t_{0,\alpha}\ ,\ v_1) = 0$, $M^{(5)}(h,t_0\backslash t_{0,\omega}) = 0$ and $M^{(5)}(h_1,t_i\backslash t_{i,\omega}) = 0$. Moreover, there are the following induced admissible transformations on submatrices of $M^{(5)}$:

(12) $(t_{p_0,\omega},v_1) \longrightarrow (t_{p_1,\omega},v_1) \longrightarrow \dots \longrightarrow (t_{0_2,\alpha},v_1) \longrightarrow (t_{0_1,\alpha},v_1) \longrightarrow$ $\longrightarrow (h_1,v_1)$.

(13) $(h_1,v_1) \longleftarrow (h_1,t_{0_1,\omega}) \longleftarrow (h_1,t_{0_2,\omega}) \longleftarrow \dots \longleftarrow (h_1,t_{i_2,\omega}) \longleftarrow$ $\longleftarrow (h_1,t_{i_1,\omega}) \longleftarrow (h_1,t_{i_0,\omega})$.

(14) $(h,t_{0,\omega}) \longleftarrow (h\ ,\ (\pi\cdot 1)|2|3|3'|4|4'|6_6|6_9)$, with change of $M^{(5)}(h,\kappa'|v')$.

(15) $$M^{(5)}\begin{pmatrix} h_2 \\ \hline h_3 \\ \hline h_4 \\ \hline h_5 \\ \hline h_7 \\ \hline h_8 \end{pmatrix}, t_{0,\omega} \sim M^{(5)}\begin{pmatrix} h_2 \\ \hline h_3 \\ \hline h_4 \\ \hline h_5 \\ \hline h_7 \\ \hline h_8 \end{pmatrix}, t_{0,\omega} + \pi^2 \begin{pmatrix} C_2R_0 - C_3 \\ \hline -C_2 + dC_3R_0 \\ \hline C_4R_0 - C_5 \\ \hline -C_4 + dC_5R_0 \\ \hline C_7R_0 - C_8 \\ \hline -C_7 + dC_8R_0 \end{pmatrix}$$, with change of $M^{(5)}(h,\kappa'|v')$, for arbitrary coefficient matrices C_i , $i = 2,3,4,5,7,8$.

(16) $(h,v_1) \longleftarrow (h\ ,\ (\pi\cdot 1)|2|3|3'|4|4'|6|6')$.

(17) $(t_0,v_9|v_6) \longleftarrow (t_0,7)$, with change of $M^{(5)}(t_0,\kappa')$.

(18) $M^{(5)}(t_0,v_8|v_7|v_5|v_4|v_3|v_2) \sim M^{(5)}(t_0,v_8|v_7|v_5|v_4|v_3|v_2) +$ $\pi^2(D_7+R_0dD_8|R_0D_7+D_8|D_4+R_0dD_5|R_0D_4+D_5|D_2+R_0dD_3|R_0D_2+D_3)$, with change of $M^{(5)}(t_0,\kappa')$, for arbitrary coefficient matrices D_i, $i = 2,3,4,5,7,8$.

(19) $(t_0,\kappa'_{0,\omega}|\kappa'_{i,\alpha}) \longleftarrow (t_0,7)$.

(20) $(h\ ,\ v'\backslash v_1) \leftarrow (h,5)$, with change of $M^{(5)}(h,\kappa')$.

(21) $(h\ ,\ \kappa'_{0,\omega}|\kappa'_{i,\alpha}) \leftarrow (h,5)$.

Applying (12) and (13) to $M_s^{(5)\prime}$ we obtain that

$$M^{(5)}\left(\frac{t}{h_1}\ ,\ v_1|t'\right) \sim$$

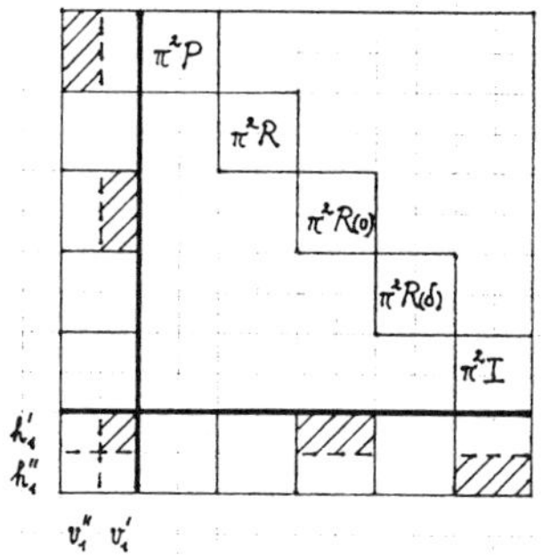

In accordance with this reduction we refine the coordinates h_1 and v_1 to $\frac{h_1'}{h_1''}$ and $v_1''|v_1'$, as indicated. In view of (14) - (16) we obtain that $M^{(5)}(h\backslash h_1\ ,\ v_1|t_0) \sim 0$. In view of (10), (17) - (21) we obtain that $M^{(5)}\left(\frac{t_0}{h_1}\ ,\ \kappa'|v'\backslash v_1\right) \sim 0$. Altogether this proves that

$$M_s^{(5)} \sim$$

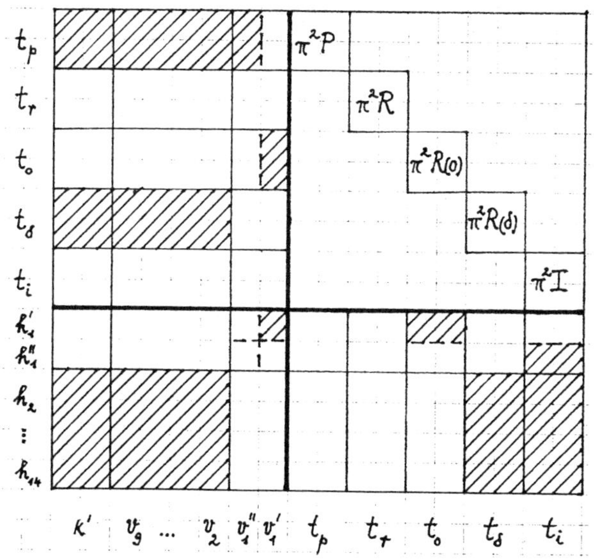

$$= M_s^{(5)\prime\prime} .$$

Now it is evident that $M^{(5)}\left(\frac{t_0}{h'_1}, v'_1|t_0\right)$ splits off from $M_s^{(5)}$. Therefore the submatrix of $M^{(5)}$ which corresponds to $M^{(5)}\left(\frac{t_0}{h'_1}, v'_1|t_0\right)$ splits off from $M^{(5)}$, and this submatrix is indeed an object in $T(0)$.

6.5. Splitting off $T(\delta)$.

We carry on with the coreduction of $M_s^{(5)''}$. Due to (1) and (6) we can assume that $M^{(5)}(t_p\backslash t_{p,\omega}, v_3|v_2) = 0$, $M^{(5)}(t_\delta\backslash t_{\delta,\omega}, v_3|v_2) = 0$, $M^{(5)}\left(\frac{h_2}{h_3}, t_\delta\backslash t_{\delta,\alpha}\right) = 0$, and $M^{(5)}\left(\frac{h_2}{h_3}, t_i\backslash t_{i,\omega}\right) = 0$. Moreover, there are the following induced admissible transformations on submatrices of $M^{(5)}$:

(22) $(t_{p_0,\omega}, v_3|v_2) \to (t_{p_1,\omega}, v_3|v_2) \to \dots \to (t_{\delta_2,\omega}, v_3|v_2) \to$
$\to (t_{\delta_1,\omega}, v_3|v_2) \to \left(\frac{h_2}{h_3}, v_3|v_2\right)$.

(23) $\left(\frac{h_2}{h_3}, v_3|v_2\right) \leftarrow \left(\frac{h_2}{h_3}, t_{\delta_1,\alpha}\right) \leftarrow \left(\frac{h_2}{h_3}, t_{\delta_2,\alpha}\right) \leftarrow \dots$
$\dots \leftarrow \left(\frac{h_2}{h_3}, t_{i_1,\omega}\right) \leftarrow \left(\frac{h_2}{h_3}, t_{i_0,\omega}\right)$.

(24) $(h, v_3|v_2) \leftarrow (h, 6_6)$, with change of $M^{(5)}\left(\begin{array}{c} h_p \\ \hline \vdots \\ \hline h_{14}\end{array}, v_3|v_2\right)$.

(25) $(h, v_3|v_2) \leftarrow (h, 2|4'|6_4)$.

(26) $(h, v_6|v_4) \leftarrow (h, 6|6')$, with change of $M^{(5)}(h,\kappa')$.

(27) $(h, \kappa'_{0,\omega}|\kappa'_{i,\alpha}|v_9|v_8|v_7|v_5) \leftarrow (h, 6|6')$.

Applying (22) and (23) to $M_s^{(5)''}$ we obtain that

$$M^{(5)}\begin{pmatrix} \frac{t}{h_2} , v_3 | v_2 | t' \\ \overline{h_3} \end{pmatrix} \sim$$

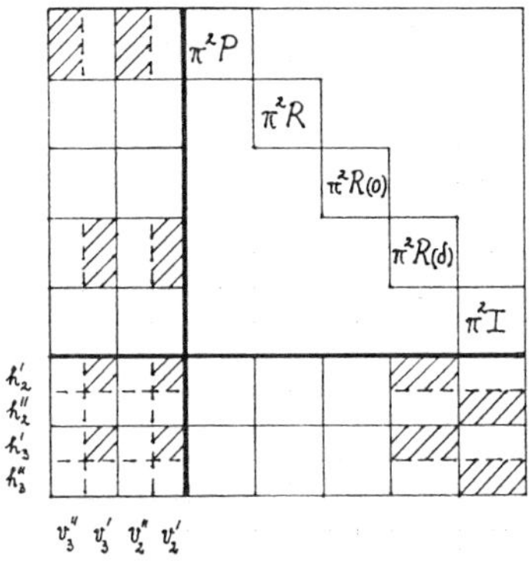

In accordance with this reduction we refine the coordinates h_j and v_j to $\frac{h'_j}{h''_j}$ and $v''_j | v'_j$, for $j = 2,3$, as indicated. In view of (2) - (5) we have $M^{(5)}(h \backslash \frac{h_2}{h_3} , t_\delta) \sim 0$. In view of (7) - (11) we have $M^{(5)}(t_\delta, \kappa' | v' \backslash (v_3 | v_2)) \sim 0$. In view of (4), (20), (24), (25) we have $M^{(5)}(h \backslash \frac{h_2}{h_3} , v_3 | v_2) \sim 0$. In view of (10), (26), (27) we have $M^{(5)}(\frac{h_2}{h_3} , \kappa' | v' \backslash (v_3 | v_2)) \sim 0$. Altogether this proves that

$$M_s^{(5)''} \sim$$

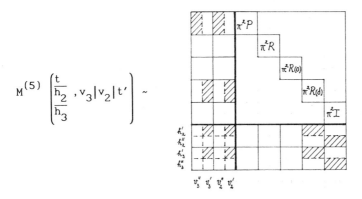

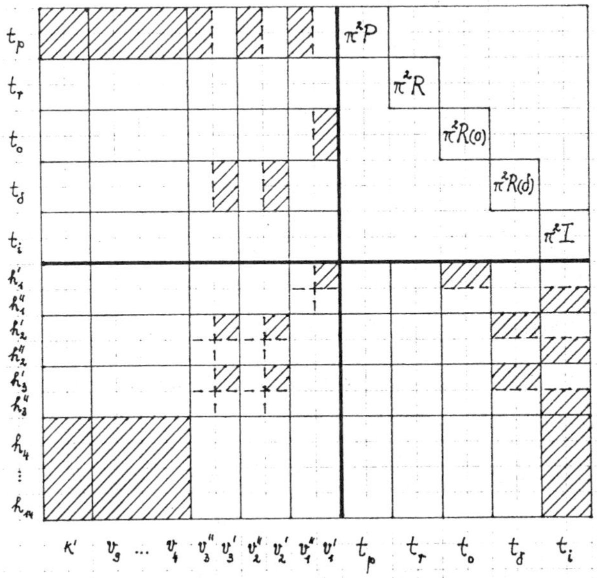

$$= M_s^{(5)'''} .$$

Now it is evident that $M^{(5)}\begin{pmatrix} \frac{t_\delta}{h'_2} \\ \overline{h'_3} \end{pmatrix}, v'_3|v'_2|t_\delta$ splits off from $M_s^{(5)}$.

Therefore the submatrix of $M^{(5)}$ which corresponds to $M^{(5)}\left(\begin{matrix} \frac{t_\delta}{h'_2} \\ \overline{h'_3} \end{matrix}, v'_3|v'_2|t_\delta\right)$ splitts off from $M^{(5)}$, and this submatrix is indeed an object in $\mathcal{T}(\delta)$.

6.6. Final transformation.

We carry on with the coreduction of $M_s^{(5)'''}$. There are the following further induced admissible transformations on submatrices of $M^{(5)}$:

(28) $(h, v_4) \leftarrow (h, 4'|2)$, with change of $M^{(5)}(h,\kappa')$.

(29) $(h, \kappa'_{0,\omega}|\kappa'_{i,\alpha}|v_9|\dots|v_5) \leftarrow (h, 4'|2)$.

In view of (4), (10), (26) - (29) we have

$$M^{(5)}\left(\begin{matrix} h_1 \\ h\backslash\overline{h_2} \\ \overline{h_3} \end{matrix}, \kappa'|v'\backslash(v_3|v_2|v_1)\right) \sim 0 .$$

Define $M_s^{(6)}$ to be the matrix which is obtained from $M_s^{(5)'''}$ by replacing $M^{(5)}\left(\begin{matrix} h_1 \\ h\backslash\overline{h_2} \\ \overline{h_3} \end{matrix}, \kappa'|v'\backslash(v_3|v_2|v_1)\right)$ by 0. Define $M^{(6)}$ to be the matrix which is obtained from $M^{(5)}$ by replacing $M_s^{(5)}$ by $M_s^{(6)}$, and keep the coordinates of $M^{(5)}$ for $M^{(6)}$. Then we have proved that $M^{(5)} \sim M^{(6)}$ and $M^{(6)} = T \oplus S$, where $T \in \mathcal{T}$ and S is the submatrix of $M^{(6)}$ which corresponds to $M^{(6)}\left(\begin{matrix} \frac{t_p}{h'_1} \\ h\backslash\overline{h'_2} \\ \overline{h'_3} \end{matrix}, \kappa'|v'\backslash(v'_3|v'_2|v'_1)|t_i\right)$.

Hence S is of the form given on the following double page. In view of step 3 the Kronecker-module corresponding to $(K|K')$ has no semisimple direct summand. In view of step 4 the k-linear map corresponding to $M^{(6)}\left(\begin{matrix}3' \\ \hline 3 \\ \hline 2\end{matrix}\ ,\ v\right)$ is injective and the k-linear map corresponding to $M^{(6)}(h,1|2|3|3'|4|4'|5'|6|6')$ is surjective. Moreover, $\pi^2 P'' \sim \begin{pmatrix}\pi^2 E \\ 0\end{pmatrix}$ by permutation of rows, and $\pi^2 I' \sim (\pi^2 E\ 0)$ by permutation of columns. Altogether this yields that S is equivalent to an object in S .

Therefore, summarizing step 1 - step 6, we have proved that $M^{(0)} \sim M^{(1)} \sim \ldots \sim M^{(6)} \in S \vee T$. q.e.d.

COROLLARY 2.2 (Second reduction). The full subcategory $\hat{S} \vee T(E)$ of $\mathcal{M}$ is dense in $\mathcal{M}$. In particular, the quivers $A(\mathcal{M})$ and $A_{\mathcal{M}}(\hat{S} \vee T(E))$ are isomorphic.

PROOF. In view of Proposition 2.1 it is sufficient to show that $\bigvee_{\lambda\in\check{I}} T(\lambda) \subset \hat{S}$. Indeed, let $T \in T(\lambda)$, for some $\lambda \in \check{I}\cdot$. Setting $(X,Y,Z) = \{X_8, X_{8'}\} = \{E, \phi_\lambda n\}$, we obtain that $(X,Y,Z) \in \hat{\mathfrak{M}}$ and $T = \sigma(X,Y,Z) \in \hat{S}$. q.e.d.

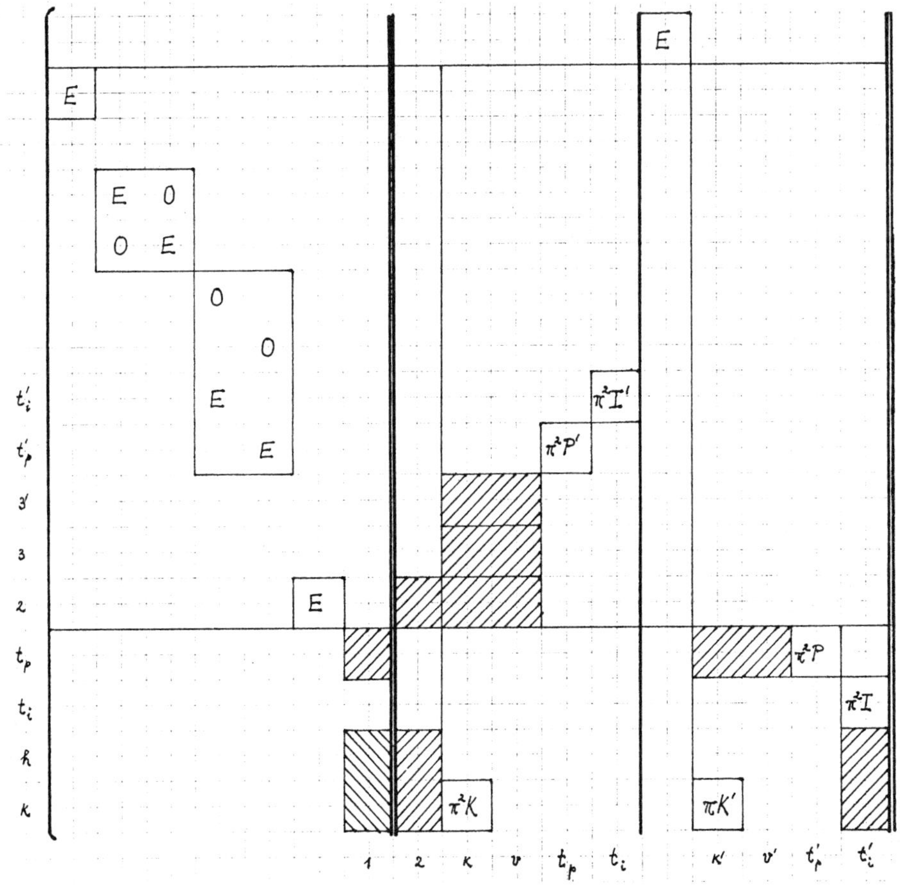
E
E
E 0
0 E
0
0
E
E
E
$\pi^2 I'$
$\pi^2 P'$
$\pi^2 K$
$\pi K'$
$\pi^2 P$
$\pi^2 I$
t_i'
t_p'
3'
3
2
t_p
t_i
h
k
1
2
k
v
t_p
t_i
k'
v'
t_p'
t_i'

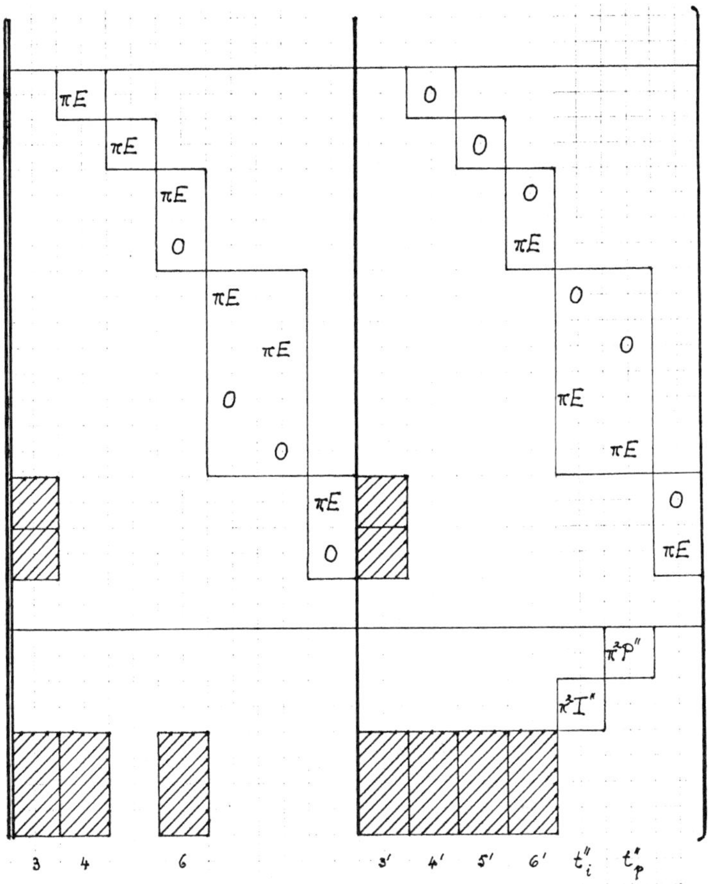
πE
πE
πE
0
πE
πE
0
0
πE
0
0
0
0
πE
0
0
πE
πE
0
πE
π³P''
π³I''
3
4
6
3'
4'
5'
6'
t''_i
t''_p

3. THIRD REDUCTION

In this third reduction we construct the quiver $A_{\mathcal{M}}(T(E))$. As a consequence, the problem of determining $A_{\mathcal{M}}(\hat{S} \vee T(E))$ will be reduced to the problem of determining $A(\hat{S})$. Although we are still working on the base-dependent level, our approach in this third reduction is in conspicuous contrast to the approach of the second reduction. Instead of reducing matrices we shall construct Auslander-Reiten components of $\mathcal{M}$.

By [Di 85] and the proof of Proposition 1.1, Auslander-Reiten sequences exist in $F(K_0)$ and their basic properties are known. In view of the representation equivalence $\phi_1 : \check{F}_0 \longrightarrow \mathcal{M}$, as defined in the proof of Proposition 1.2, these carry over to $\mathcal{M}$. As a consequence we obtain the following basic facts.

(1) Each Auslander-Reiten sequence in $\mathcal{M}$ is uniquely determined, up to isomorphism, by its starting term as well as by its end term.

(2) If $M \in \mathcal{M}$ is indecomposable and not isomorphic to one of the objects $(K_1, R_3, (0|1))$, $(K_i, 0, 0)$, $i = 0,1,2$, then there exists an Auslander-Reiten sequence starting in M . If $M \in \mathcal{M}$ is indecomposable and not isomorphic to one of the objects $(K_0, R_3, (1))$, $(0, R_3, 0)$, $(K_1, R_3^2, \begin{pmatrix} \pi^2 & 0 \\ 0 & 1 \end{pmatrix})$, $(K_2, R_3^2, \begin{pmatrix} \pi & 0 \\ 0 & \pi \end{pmatrix})$, then there exists an Auslander--Reiten sequence ending in M .

(3) All components of a morphism which belongs to an Auslander-Reiten sequence are irreducible morphisms. Conversely, each irreducible morphism between indecomposable objects is a component of a morphism which belongs to an Auslander-Reiten sequence.

For details we refer to [Di 85]. (Note that (1) and (3) are general statements on Auslander-Reiten sequences, independent of the category $\mathcal{M}$.)

For any morphism φ in $\mathcal{M}$ we write $\varphi = (\tau(\varphi),\sigma(\varphi))$ if there is need to emphasize the fact that each morphism in $\mathcal{M}$ consists of a pair of morphisms in $\mathrm{mod}R_3$.

The construction of $A_{\mathcal{M}}(T(E))$ which we are aiming at will be based on the following two Lemmata.

LEMMA 3.1. Let $M' \xrightarrow{\varphi} E \xrightarrow{\psi} M$ be an Auslander-Reiten sequence in $\mathcal{M}$ such that in the Krull-Schmidt decomposition $E = \bigoplus_{i=1}^{n} E_i$ of E the indecomposable direct summands E_i are pairwise non-isomorphic. Moreover, let $\mathfrak{E} : M' \xrightarrow{\alpha} X \xrightarrow{\beta} M$ be a non-split short exact sequence in $\mathcal{M}$ and let $X = \bigoplus_{i=1}^{m} X_i$ be a decomposition of X into pairwise non-isomorphic direct summands X_i such that X_1 is indecomposable and the components $\alpha_1 : M' \to X_1$ of α and $\beta_1 : X_1 \to M$ of β are irreducible. Then $\mathfrak{E}$ is an Auslander-Reiten sequence.

PROOF. Put $\varphi = (\varphi_i)_{i=1,\dots,n}$, $\psi = (\psi_i)_{i=1,\dots,n}$ and $\alpha = (\alpha_i)_{i=1,\dots,m}$, $\beta = (\beta_i)_{i=1,\dots,m}$, corresponding to the decompositions of E and X . We may assume that $E_1 = X_1$ and $\varphi_1 = \alpha_1$, $\psi_1 = \beta_1$. There exist morphisms $\gamma : X \to E$ and $\delta : M' \to M'$ such that $\psi\gamma = \beta$ and $\varphi\delta = \gamma\alpha$. Put $\gamma = (\gamma_{ij})_{\substack{i=1,\dots,n \\ j=1,\dots,m}}$, according to the decompositions of E and X . Then $\psi_1 = \beta_1 = \psi_1\gamma_{11} + \sum_{i=2}^{n} \psi_i\gamma_{i1} \equiv \psi_1\gamma_{11}$ $(\mathrm{mod}\ \mathrm{rad}^2_{\mathcal{M}}(E_1,M))$. Hence $\gamma_{11} : X_1 \to E_1$ is an isomorphism. Moreover, $\varphi_1\delta = \gamma_{11}\alpha_1 + \sum_{i=2}^{m} \gamma_{1i}\alpha_i \equiv \gamma_{11}\varphi_1$ $(\mathrm{mod}\ \mathrm{rad}^2_{\mathcal{M}}(M',E_1))$. Hence δ is an isomorphism. Consequently γ is an isomorphism. Therefore $\mathfrak{E}$ is isomorphic to the Auslander-Reiten sequence $M' \xrightarrow{\varphi} E \xrightarrow{\psi} M$. Hence $\mathfrak{E}$ is an Auslander-Reiten sequence itself. q.e.d.

LEMMA 3.2. Suppose we are given in $\mathcal{M}$ a set of objects $\{M_{rn} \mid r \in \mathbb{Z}/2\mathbb{Z},\ n \in \mathbb{N}\}$ and a set of morphisms $\{\iota_{rn} : M_{r,n-1} \longrightarrow M_{rn}$, $\pi_{rn} : M_{rn} \longrightarrow M_{r+1,n-1} \mid r \in \mathbb{Z}/2\mathbb{Z}$, $n \geq 2\}$, subject to the following conditions:

(c_1) $\mathfrak{E}_{11} : M_{21} \xrightarrow{\iota_{22}} M_{22} \xrightarrow{\pi_{22}} M_{11}$ and $\mathfrak{E}_{21} : M_{11} \xrightarrow{\iota_{12}} M_{12} \xrightarrow{\pi_{12}} M_{21}$ are Auslander-Reiten sequences in $\mathcal{M}$ such that M_{11} and M_{21} are non--isomorphic and M_{12} and M_{22} are indecomposable.

(c_2) $\underline{\dim}\, M_{rn} + \underline{\dim}\, M_{r+1,n} = \underline{\dim}\, M_{r+1,n-1} + \underline{\dim}\, M_{r,n+1}$, for all $r \in \mathbb{Z}/2\mathbb{Z}$, $n \geq 2$.

(c_3) $\underline{\dim}\, M_{rn}$, $\underline{\dim}\, M_{r+1,n-1}$ and $\underline{\dim}\, M_{r,n+1}$ are pairwise different, for all $r \in \mathbb{Z}/2\mathbb{Z}$, $n \geq 2$.

(c_4) $\tau(\pi_{rn})$ and $\sigma(\pi_{rn})$ are splittable epimorphisms, for all $r \in \mathbb{Z}/2\mathbb{Z}$, $n \geq 2$.

(c_5) $\tau(\iota_{rn})$ and $\sigma(\iota_{rn})$ are splittable monomorphisms, for all $r \in \mathbb{Z}/2\mathbb{Z}$, $n \geq 2$.

(c_6) $\iota_{r+1,n}\pi_{rn} = \pi_{r,n+1}\iota_{r,n+1}$, for all $r \in \mathbb{Z}/2\mathbb{Z}$, $n \geq 2$.

Then all objects in $\{M_{rn} \mid r \in \mathbb{Z}/2\mathbb{Z}$, $n \in \mathbb{N}\}$ are indecomposable and pairwise nonisomorphic. They constitute the set of points of a connected component of $A(\mathcal{M})$ which has the structure of a stable tube of rank 2. Setting $m_{rn} = [M_{rn}]$, it is given as follows.

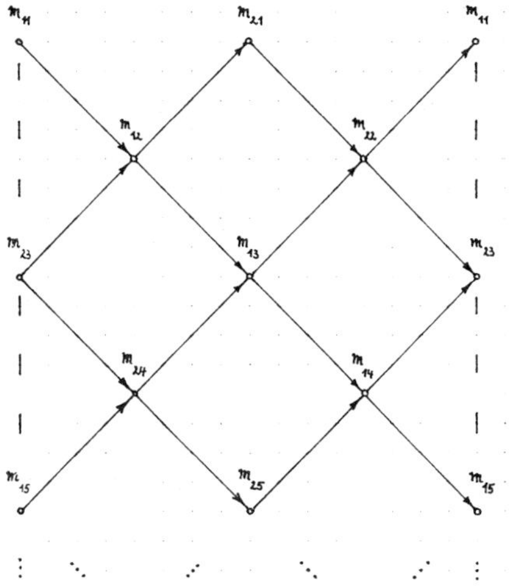

PROOF. Put $\mathfrak{E}_{r+1,n} : M_{rn} \xrightarrow{\begin{pmatrix}\pi_{rn}\\ \iota_{r,n+1}\end{pmatrix}} M_{r+1,n-1} \oplus M_{r,n+1} \xrightarrow{(\iota_{r+1,n}\ -\pi_{r,n+1})} M_{r+1,n}$, for all $r \in \mathbb{Z}/2\mathbb{Z}$, $n \geq 2$. By (c_1) we know that $\tau[M_{11}] \neq [M_{11}]$ and $\tau^2[M_{11}] = [M_{11}]$. Therefore the component of $A(\mathcal{M})$ which contains $[M_{11}]$ is a stable tube of rank 2. Hence it suffices to prove that all sequences $\mathfrak{E}_{rn}$ are Auslander-Reiten sequences in $\mathcal{M}$.

First of all we show that all sequences $\mathfrak{E}_{rn}$ are short exact. Under the equivalence of categories $F(K_1) \cong \mathcal{M}$ each object M_{rn} of $\mathcal{M}$ corresponds to an object $(K_{rn}, F_{rn}, M_{rn} : K_{rn} \longrightarrow F_{rn})$ of $F(K_1)$. Now let $n \geq 2$ and put $\alpha = \begin{pmatrix}\pi_{rn}\\ \iota_{r,n+1}\end{pmatrix}$ and $\beta = (\iota_{r+1,n}\ -\pi_{r,n+1})$. Associated with $\mathfrak{E}_{r+1,n}$ there are the sequences

$\tau(\mathfrak{E}_{r+1,n}) : K_{rn} \xrightarrow{\tau(\alpha)} K_{r+1,n-1} \oplus K_{r,n+1} \xrightarrow{\tau(\beta)} K_{r+1,n}$ in K_1 and

$\sigma(\mathfrak{E}_{r+1,n}) : F_{rn} \xrightarrow{\sigma(\alpha)} F_{r+1,n-1} \oplus F_{r,n+1} \xrightarrow{\sigma(\beta)} F_{r+1,n}$ in $\mathrm{mod}R_3$.

We have to show that they are both split short exact. By (c_2) and (c_4) we have $K_{rn} \oplus K_{r+1,n} \cong K_{r+1,n-1} \oplus K_{r,n+1} \cong \ker\tau(\beta) \oplus K_{r+1,n}$, hence

$K_{rn} \cong \ker\tau(\beta)$. By (c_5) and (c_6) we have $K_{rn} \cong \operatorname{im}\tau(\alpha) \subset \ker\tau(\beta)$, hence $\operatorname{im}\tau(\alpha) = \ker\tau(\beta)$. Therefore $\tau(\mathfrak{E}_{r+1,n})$ is split short exact. Analogously one verifies that $\sigma(\mathfrak{E}_{r+1,n})$ is split short exact. Thus $\mathfrak{E}_{r+1,n}$ is a short exact sequence in $\mathcal{M}$.

Now we prove by induction on n that all sequences $\mathfrak{E}_{rn}$ are Auslander-Reiten sequences in $\mathcal{M}$. For $n = 1$ this is true by hypothesis (c_1). Let $n \geq 2$ and suppose that $\mathfrak{E}_{r,n-1}$ is an Auslander-Reiten sequence in $\mathcal{M}$, for all $r \in \mathbb{Z}/2\mathbb{Z}$. As we have shown above, $\mathfrak{E}_{r+1,n}$ is a short exact sequence. By hypothesis, $M_{r+1,n-1}$ is indecomposable and π_{rn} and $\iota_{r+1,n}$ are irreducible morphisms. By (c_3), $\mathfrak{E}_{r+1,n}$ is non-split and the objects $M_{r+1,n-1}$ and $M_{r,n+1}$ are not isomorphic. Hence we may apply Lemma 3.1 which yields the desired result that $\mathfrak{E}_{r+1,n}$ is an Auslander-Reiten sequence. q.e.d.

We proceed towards the construction of the full subquiver $A_{\mathcal{M}}(T(E))$ of $A(\mathcal{M})$. As a preparatory step we present a list of distinguished objects $\{T(\varepsilon)_n, T(\bar{\varepsilon})_n \mid n \in \mathbb{N}\}$ which are isomorphic to objects in $T(\varepsilon)$, for each $\varepsilon \in E$. In this context, m will denote an arbitrary nonzero natural number.

1) <u>List of objects</u> $T(\infty)_n$ <u>and</u> $T(\bar{\infty})_n$, $n \in \mathbb{N}$.

$$T(\infty)_n = \left(\begin{array}{c||c|c||c|c} 0 & & & \pi E & 0 \\ E & \pi^2 R(\infty)_n' & & 0 & \pi E \\ \hline & & \pi^2 R(\infty)_n & & \pi^2 R(\infty)_n'' \end{array}\right) , \quad T(\bar{\infty})_n = \left(\begin{array}{c||c|c||c|c} 0 & & & \pi E & 0 \\ E & \pi^2 R(\bar{\infty})_n' & & 0 & \pi E \\ \hline & & \pi^2 R(\bar{\infty})_n & & \pi^2 R(\bar{\infty})_n'' \end{array}\right)$$

where the triples of k-matrices $(R(\infty)_n, R(\infty)_n', R(\infty)_n'')$ and $(R(\bar{\infty})_n, R(\bar{\infty})_n', R(\bar{\infty})_n'')$ are the normal forms of the indecomposable objects in

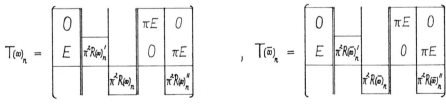

$\text{reg}_\infty kQ_3$, as defined in 0.3.

2) <u>List of objects</u> $T(0)_n$ <u>and</u> $T(\bar{0})_n$, $n \in \mathbb{N}$.

$T(0)_1 = (1 \| \pi^2 | 0)$, with <u>dim</u> $T(0)_1 = (1;1,1,0)$,

$T(0)_{2m} =$ 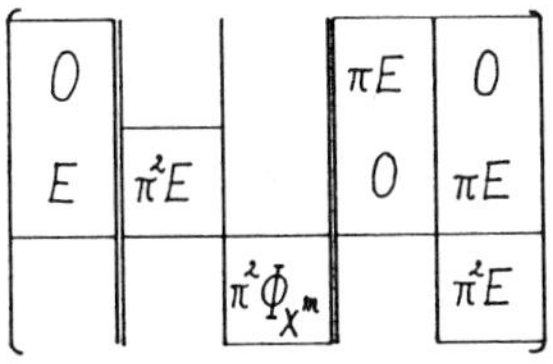 , $T(0)_{2m+1} =$ 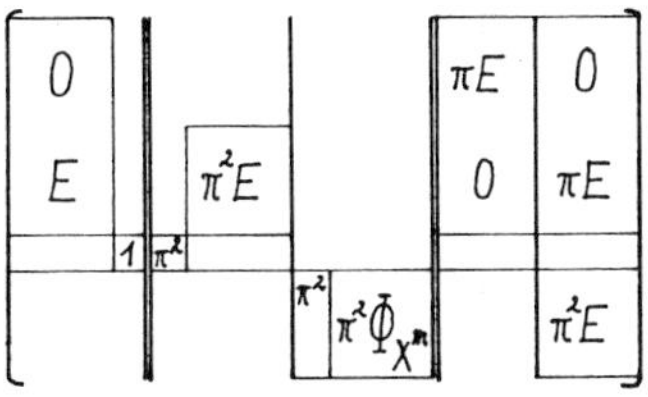

$T(\bar{0})_1 = \left(\begin{array}{c|c} \pi & 0 \\ 0 & \pi^2 \end{array}\right)$, with <u>dim</u> $T(\bar{0})_1 = (2;,0,0,1)$,

$$T(\bar{0})_2 = \left[\begin{array}{c|c|c|c|c} 0 & 0 & 0 & \pi & 0 \\ 1 & \pi^2 & 0 & 0 & 0 \\ 0 & 0 & \pi^2 & 0 & \pi^2 \end{array}\right]$$

$T(\bar{0})_{2m+2} =$ 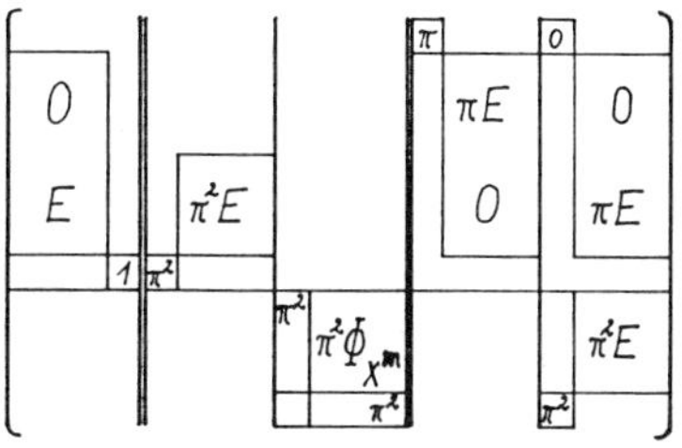 , $T(\bar{0})_{2m+1} =$ 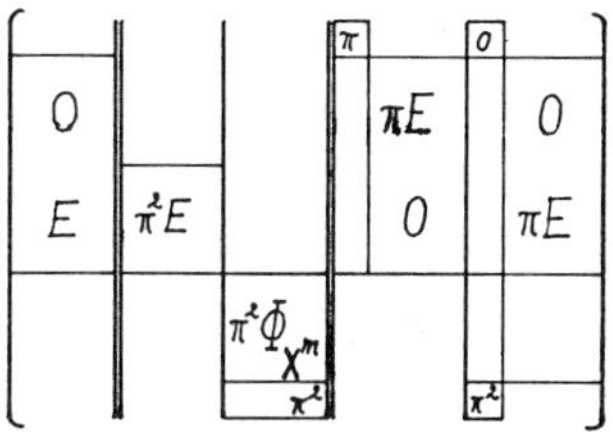

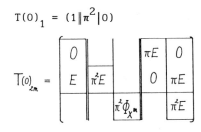

3) <u>List of objects</u> $T(\delta)_n$ <u>and</u> $T(\bar{\delta})_n$, $n \in \mathbb{N}$.

$T(\delta)_1 = \left[\begin{array}{cc|cc|c|c} \pi^2 & 0 & 0 & 0 & \pi & 0 \\ 0 & \pi^2 & 0 & 0 & 0 & \pi \end{array}\right]$, with <u>dim</u> $T(\delta)_1 = (2;0,2,1)$,

$$T(\delta)_{2m} = \left(\begin{array}{cc||cc|cc||c|c|c|c}
 & & \pi^2 E & & 0 & & \pi E & & 0 & \\
 & & & \pi^2 E & & 0 & 0 & \pi E & \pi E & \\
\hline
E & & & & & & & \pi E & & 0 \\
 & E & & & & & & 0 & & \pi E \\
 & & & & \pi^2 \Phi_{\chi^m} & & & \pi^2 E & & 0 \\
 & & & & & & & 0 & & \pi^2 E
\end{array}\right)$$

$$T(\delta)_{2m+1} = \left(\begin{array}{cc||cc|cc||c|c|c|c}
 & & \pi^2 E_{m+1} & & 0 & & \pi E_{m+1} & & 0 & \\
 & & & & & & & 0 \ldots 0 & & \\
 & & & \pi^2 E_{m+1} & & 0 & 0 & \pi E_m & \pi E_{m+1} & \\
\hline
E_m & & & & & & & \pi E_m & & 0 \\
 & E_m & & & & & & 0 & & \pi E_m \\
 & & & & \pi^2 E_m \begin{smallmatrix} 0 \\ \vdots \\ 0 \end{smallmatrix} & & & \pi^2 E_m & & 0 \\
 & & & & & & & 0 & & \pi^2 E_m
\end{array}\right)$$

$$T(\bar{\delta})_1 = \left(\begin{array}{cc||c|c}
1 & 0 & \pi & 0 \\
0 & 1 & 0 & \pi \\
0 & 0 & \pi^2 & 0 \\
0 & 0 & 0 & \pi^2
\end{array}\right) \quad , \text{ with } \underline{\dim}\ T(\bar{\delta})_1 = (4;2,0,1) ,$$

$$T(\bar{\delta})_{2m} = \left(\begin{array}{cc||cc|cc||c|c|c|c}
 & & \pi^2 E & & 0 & & \pi E & & 0 & \\
 & & & \pi^2 E & & 0 & 0 & \pi \Phi_{\chi^m} & \pi E & \\
\hline
E & & & & & & & \pi E & & 0 \\
 & E & & & & & & 0 & & \pi E \\
 & & & & \pi^2 E & & & \pi^2 E & & 0 \\
 & & & & & & & 0 & & \pi^2 E
\end{array}\right)$$

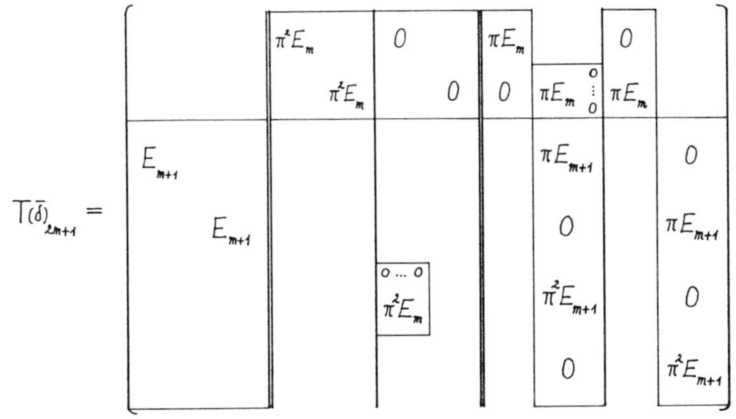

LEMMA 3.3. Every object in $T(\varepsilon)$ is isomorphic to an object in $\mathrm{add}\{T(\varepsilon)_n, T(\bar{\varepsilon})_n \mid n \in \mathbb{N}\}$, for each $\varepsilon \in E$.

PROOF. For $\varepsilon = \infty$, this is true by definition of $T(\infty)$. For $\varepsilon = 0$, consider the matrix problem which is induced on the submatrix

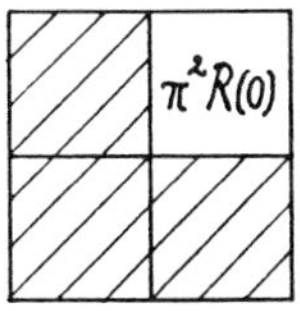

of $T(0)$. It is an exercise to show that any object of the induced matrix problem decomposes into direct summands of the following type.

I , ⊢⊣ , (π^2) , where $\pi^2 R(0) = \emptyset$;

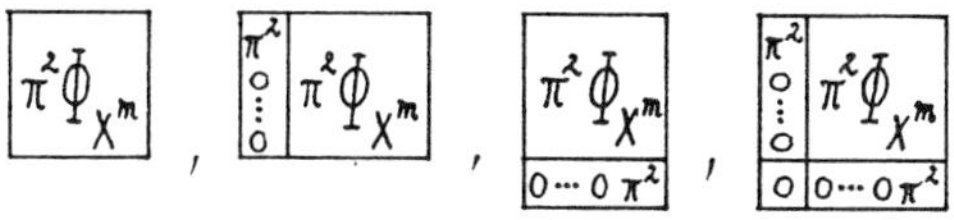

, where $m \in \mathbb{N}$.

The objects in $T(0)$ which correspond to these direct summands are just the objects in $\{T(0)_n, T(\bar{0})_n \mid n \in \mathbb{N}\}$. For $\varepsilon = \delta$, consider the matrix problem which is induced on the submatrix

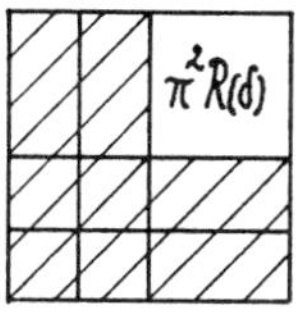

of $T(\delta)$. Again it is an exercise to show that any object of the induced matrix problem decomposes into direct summands of the following type.

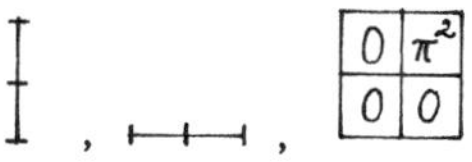

, where $\pi^2 R(\delta) = \emptyset$;

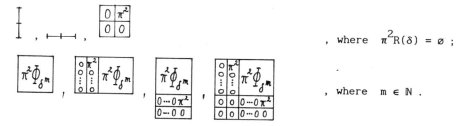

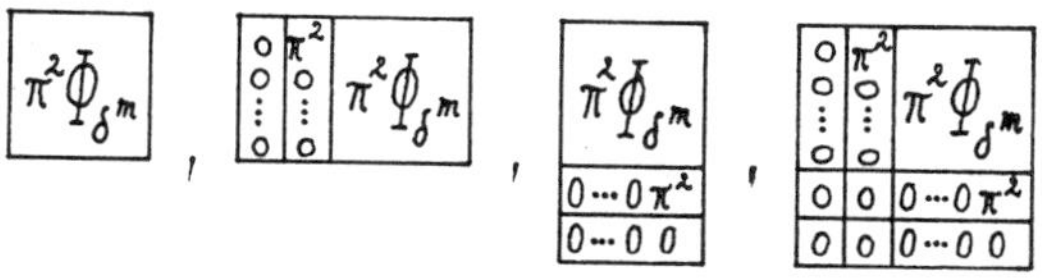

, where $m \in \mathbb{N}$.

The objects in $T(\delta)$ which correspond to these direct summands are isomorphic to objects in $\{T(\delta)_n, T(\bar{\delta})_n \mid n \in \mathbb{N}\}$. The finding of these isomorphisms is once more an exercise in matrix problems. q.e.d.

PROPOSITION 3.4. For each $\varepsilon \in E$ we have:

(i) The full subquiver $A_{\mathcal{M}}(T(\varepsilon))$ of $A(\mathcal{M})$ is a connected component of $A(\mathcal{M})$ which has the structure of a stable tube of rank 2.

(ii) All objects in $\{T(\varepsilon)_n, T(\bar{\varepsilon})_n \mid n \in \mathbb{N}\}$ are indecomposable and pairwise non-isomorphic. They constitute the set of points of $A_{\mathcal{M}}(T(\varepsilon))$.

(iii) Setting $t(\varepsilon)_n = [T(\varepsilon)_n]$ and $t(\bar{\varepsilon})_n = [T(\bar{\varepsilon})_n]$, the component $A_{\mathcal{M}}(T(\varepsilon))$ is given as follows.

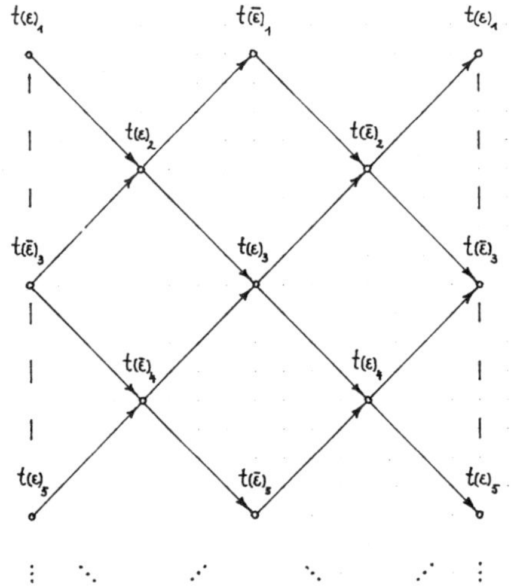

PROOF. We introduce an abbreviating notation for matrices, for the purposes of this proof only. $\Delta(M_1,\dots,M_n)$ denotes the diagonal block-matrix $(\delta_{ij}M_i)_{i,j=1,\dots,n}$. In $R_3^{m\times m}$ we specify the two matrices

$$E_m = \begin{pmatrix} 1 & & \\ & \ddots & \\ & & 1 \end{pmatrix} \quad \text{and} \quad I_m = \begin{pmatrix} & & 1 \\ & \cdot^{\cdot^{\cdot}} & \\ 1 & & \end{pmatrix}.$$

With any fixed matrix $M \in R_3^{m\times n}$ we associate matrices in $R_3^{(m+1)\times n}$, in $R_3^{m\times(n+1)}$, and in $R_3^{(m+1)\times(n+1)}$ as follows: $\overset{\circ}{M} = \begin{pmatrix} 0\dots0 \\ M \end{pmatrix}$, $\underset{\circ}{M} = \begin{pmatrix} M \\ 0\dots0 \end{pmatrix}$, $M\circ = \begin{pmatrix} M & \begin{matrix} 0 \\ \vdots \\ 0 \end{matrix} \end{pmatrix}$,

$$\circ M = \begin{pmatrix} \begin{matrix} 0 \\ \vdots \\ 0 \end{matrix} & M \end{pmatrix}, \quad M^{\urcorner} = \begin{pmatrix} 0\dots0 & 1 \\ M & \begin{matrix} 0 \\ \vdots \\ 0 \end{matrix} \end{pmatrix}, \quad {}^{\ulcorner}M = \begin{pmatrix} 1 & 0\dots0 \\ \begin{matrix} 0 \\ \vdots \\ 0 \end{matrix} & M \end{pmatrix}, \quad M_{\lrcorner} = \begin{pmatrix} M & \begin{matrix} 0 \\ \vdots \\ 0 \end{matrix} \\ 0\dots0 & 1 \end{pmatrix}.$$

Recall that any morphism μ in $\mathcal{M}$ is given by a pair $(\tau(\mu),\sigma(\mu))$. Throughout this proof m denotes a natural number.

With these conventions in mind we now introduce a set of morphisms $\{\iota_n : T(\varepsilon)_{n-1} \longrightarrow T(\varepsilon)_n$, $\bar{\iota}_n : T(\bar{\varepsilon})_{n-1} \longrightarrow T(\bar{\varepsilon})_n$, $\pi_n : T(\varepsilon)_n \longrightarrow T(\bar{\varepsilon})_{n-1}$, $\bar{\pi}_n : T(\bar{\varepsilon})_n \longrightarrow T(\varepsilon)_{n-1}$, $n \geq 2\}$ in $T(\varepsilon)$, for each $\varepsilon \in E$.

1) Definition of morphisms in $T(\infty)$.

$\tau(\iota_{2m}) = \Delta(I_m, \mathring{I}_{m-1}, \mathring{I}_{m-1}, I_m, I_m)$, $\sigma(\iota_{2m}) = \Delta(I_m, I_m, \mathring{I}_{m-1})$

$\tau(\iota_{2m+1}) = \Delta(\underset{\circ}{I}_m, I_m, I_m, \underset{\circ}{I}_m, \underset{\circ}{I}_m)$, $\sigma(\iota_{2m+1}) = \Delta(\underset{\circ}{I}_m, \underset{\circ}{I}_m, I_m)$

$\tau(\bar{\iota}_{2m}) = \Delta(\mathring{E}_{m-1}, E_m, E_m, \mathring{E}_{m-1}, \mathring{E}_{m-1})$, $\sigma(\bar{\iota}_{2m}) = \Delta(\mathring{E}_{m-1}, \mathring{E}_{m-1}, E_m)$

$\tau(\bar{\iota}_{2m+1}) = \Delta(E_m, \mathring{E}_m, \mathring{E}_m, E_m, E_m)$, $\sigma(\bar{\iota}_{2m+1}) = \Delta(E_m, E_m, \mathring{E}_m)$

$\tau(\pi_{2m}) = \Delta(E^{\circ}_{m-1}, E_m, E_m, E^{\circ}_{m-1}, E^{\circ}_{m-1})$, $\sigma(\pi_{2m}) = \Delta(E^{\circ}_{m-1}, E^{\circ}_{m-1}, E_m)$

$\tau(\pi_{2m+1}) = \Delta({}^{\circ}I_m, I_m, I_m, {}^{\circ}I_m, {}^{\circ}I_m)$, $\sigma(\pi_{2m+1}) = \Delta({}^{\circ}I_m, {}^{\circ}I_m, I_m)$

$\tau(\bar{\pi}_{2m}) = \Delta(I_m, I^{\circ}_{m-1}, I^{\circ}_{m-1}, I_m, I_m)$, $\sigma(\bar{\pi}_{2m}) = \Delta(I_m, I_m, I^{\circ}_{m-1})$

$\tau(\bar{\pi}_{2m+1}) = \Delta(E_m, E^{\circ}_m, E^{\circ}_m, E_m, E_m)$, $\sigma(\bar{\pi}_{2m+1}) = \Delta(E_m, E_m, E^{\circ}_m)$

2) Definition of morphisms in $T(0)$.

$\tau(\iota_2) =$ 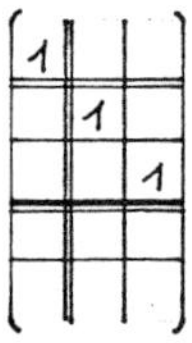 , $\sigma(\iota_2) = \begin{bmatrix} 0 \\ 1 \\ 0 \end{bmatrix}$

$\tau(\iota_{2m+1}) = \Delta(\underset{\circ}{E}_m, \mathring{E}_m, \mathring{E}_m, E_m, E_m)$, $\sigma(\iota_{2m+1}) = \Delta(E_m, \underset{\circ}{E}_m, E_m)$

$\tau(\iota_{2m+2}) = \Delta(E^{\urcorner}_m, E_{m+1}, E_{m+1}, \mathring{E}_m, \mathring{E}_m)$, $\sigma(\iota_{2m+2}) = \Delta(\mathring{E}_m, E^{\urcorner}_m, \mathring{E}_m)$

$\tau(\bar{\iota}_2) =$ 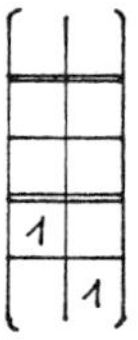 , $\sigma(\bar{\iota}_2) = \begin{bmatrix} 1 & 0 \\ 0 & 0 \\ 0 & 1 \end{bmatrix}$

$\tau(\bar{\iota}_3) = \Delta(1, 1, 1, \begin{pmatrix} 1 \\ 0 \end{pmatrix}, \begin{pmatrix} 1 \\ 0 \end{pmatrix})$, $\sigma(\bar{\iota}_3) = \Delta(\begin{pmatrix} 1 \\ 0 \end{pmatrix}, 1, \begin{pmatrix} 0 \\ 1 \end{pmatrix})$

$\tau(\bar{\iota}_{2m+2}) = \Delta(\underset{\circ}{E}_m, \mathring{E}_m, \mathring{E}_m, E_{m+1}, E_{m+1})$, $\sigma(\bar{\iota}_{2m+2}) = \Delta(E_{m+1}, \underset{\circ}{E}_m, E_{m+1})$

$\tau(\bar{\iota}_{2m+3}) = \Delta(E^{\urcorner}_m, E_{m+1}, E_{m+1}, {}^{\ulcorner}\mathring{E}_m, {}^{\ulcorner}\mathring{E}_m)$, $\sigma(\bar{\iota}_{2m+3}) = \Delta({}^{\ulcorner}\mathring{E}_m, E^{\urcorner}_m, \mathring{E}_{m+1})$

$\tau(\pi_2) =$ 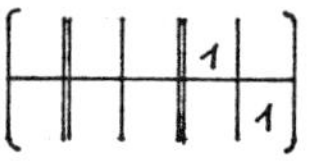 , $\sigma(\pi_2) = \begin{bmatrix} 1 & 0 & 0 \\ 0 & 0 & 1 \end{bmatrix}$

$\tau(\pi_3) = \Delta((01), (10), (10), 1, 1)$, $\sigma(\pi_3) = \Delta(1, (01), 1)$

$$\tau(\pi_{2m+2}) = \Delta(E^{\circ}_{m}, E^{\circ}_{m}, E^{\circ}_{m}, E^{\rceil}_{m}, E^{\rceil}_{m}) \quad , \quad \sigma(\pi_{2m+2}) = \Delta(E^{\rceil}_{m}, E^{\circ}_{m}, E_{m+1})$$

$$\tau(\pi_{2m+3}) = \Delta(E^{\circ}_{m\rfloor}, E^{\circ}_{m+1}, E^{\circ}_{m+1}, E^{\rceil}_{m}, E^{\rceil}_{m}) \quad , \quad \sigma(\pi_{2m+3}) = \Delta(E^{\rceil}_{m}, E^{\circ}_{m\rfloor}, E_{m+1})$$

$\tau(\bar{\pi}_2) =$ 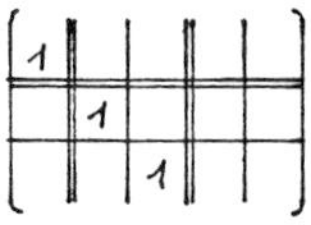 , $\sigma(\bar{\pi}_2) = (010)$

$$\tau(\bar{\pi}_{2m+1}) = \Delta(E_m, E_m, E_m, {\circ}E_m, {\circ}E_m) \quad , \quad \sigma(\bar{\pi}_{2m+1}) = \Delta({\circ}E_m, E_m, E^{\circ}_{m})$$

$$\tau(\bar{\pi}_{2m+2}) = \Delta(E_{m+1}, E_{m+1}, E_{m+1}, {\circ}E_m, {\circ}E_m) \quad , \quad \sigma(\bar{\pi}_{2m+2}) = \Delta({\circ}E_m, E_{m+1}, E^{\circ}_{m}) \ .$$

3) Definition of morphisms in $T(\delta)$.

$\tau(\iota_2) =$ 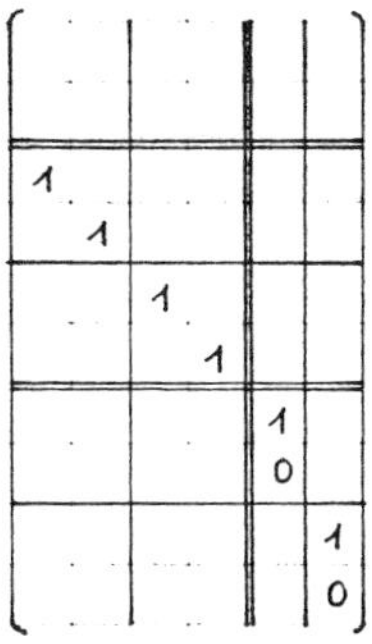 , $\sigma(\iota_2) =$ 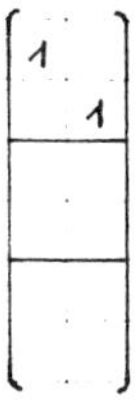

$$\tau(\iota_{2m+1}) = \Delta(E_{2m}, \mathring{\hat{E}}_m, \mathring{\hat{E}}_m, \mathring{\hat{E}}_m, \mathring{\hat{E}}_m, \mathring{\hat{E}}_m, E_m, \mathring{\hat{E}}_m, E_m) \quad , \quad \sigma(\iota_{2m+1}) = \Delta(E_m, E_m, E_{4m})$$

$$\tau(\iota_{2m+2}) = \Delta(\mathring{\hat{E}}_m, \mathring{\hat{E}}_m, E_{2m+2}, E_{2m+2}, E_{m+1}, \mathring{\hat{E}}_m, E_{m+1}, \mathring{\hat{E}}_m) \quad , \quad \sigma(\iota_{2m+2}) = \Delta(E_{2m+2}, \mathring{\hat{E}}_m, \mathring{\hat{E}}_m, \mathring{\hat{E}}_m, \mathring{\hat{E}}_m)$$

$\tau(\bar{\iota}_2) =$ 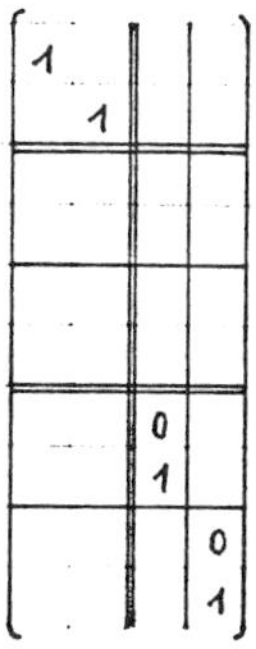 , $\sigma(\bar{\iota}_2) =$ 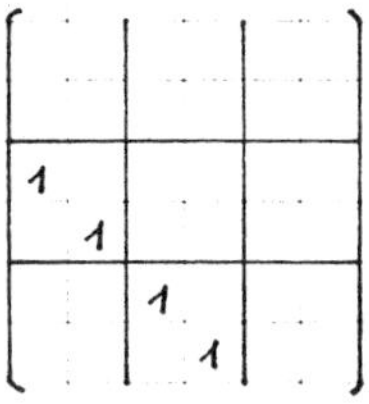

$$\tau(\bar{\iota}_{2m+1}) = \Delta(\mathring{\hat{E}}_m, \mathring{\hat{E}}_m, E_{2m}, E_{2m}, E_m, \mathring{\hat{E}}_m, E_m, \mathring{\hat{E}}_m) \quad , \quad \sigma(\bar{\iota}_{2m+1}) = \Delta(E_{2m}, E_m, E_m, E_m, E_m)$$

$$\tau(\bar{\iota}_{2m+2}) = \Delta(E_{2m+2},\mathring{E}_m,\mathring{E}_m,\mathring{E}_m,\mathring{E}_m,\mathring{E}_m,E_{m+1},\mathring{E}_m,E_{m+1}) \quad , \quad \sigma(\bar{\iota}_{2m+2}) = \Delta(\mathring{E}_m,\mathring{E}_m,E_{4m+4})$$

$$\tau(\pi_2) = \left[\begin{array}{c||c|c||c|c} \begin{smallmatrix}1 & \\ & 1\end{smallmatrix} & & & & \\ \hline\hline & & & 0\ 1 & \\ \hline & & & & 0\ 1 \end{array}\right] \quad , \quad \sigma(\pi_2) = \left[\begin{array}{c|c|c} & \begin{smallmatrix}1 & \\ & 1\end{smallmatrix} & \\ \hline & & \begin{smallmatrix}1 & \\ & 1\end{smallmatrix} \end{array}\right]$$

$$\tau(\pi_{2m+1}) = \Delta(E_{2m},E_{\mathring{m}},E_{\mathring{m}},E_{\mathring{m}},E_{\mathring{m}},E_{\mathring{m}},E_m,E_{\mathring{m}},E_m) \quad , \quad \sigma(\pi_{2m+1}) = \Delta(E_{\mathring{m}},E_{\mathring{m}},E_{4m})$$

$$\tau(\pi_{2m+2}) = \Delta(E_{2m+2},E_{\mathring{m}},E_{\mathring{m}},E_{\mathring{m}},E_{\mathring{m}},E_{\mathring{m}},E_{m+1},E_{\mathring{m}},E_{m+1}) \quad , \quad \sigma(\pi_{2m+2}) = \Delta(E_{\mathring{m}},E_{\mathring{m}},E_{4m+4})$$

$$\tau(\bar{\pi}_2) = \left[\begin{array}{c||c|c||c|c} & \begin{smallmatrix}1 & \\ & 1\end{smallmatrix} & & & \\ \hline & & \begin{smallmatrix}1 & \\ & 1\end{smallmatrix} & & \\ \hline\hline & & & 1\ 0 & \\ \hline & & & & 1\ 0 \end{array}\right] \quad , \quad \sigma(\bar{\pi}_2) = \left[\begin{array}{c|c|c} \begin{smallmatrix}1 & \\ & 1\end{smallmatrix} & & \end{array}\right]$$

$$\tau(\bar{\pi}_{2m+1}) = \Delta(E_{\mathring{m}},E_{\mathring{m}},E_{2m},E_{2m},E_m,E_{\mathring{m}},E_m,E_{\mathring{m}}) \quad , \quad \sigma(\bar{\pi}_{2m+1}) = \Delta(E_{2m},E_{\mathring{m}},E_{\mathring{m}},E_{\mathring{m}},E_{\mathring{m}})$$

$$\tau(\bar{\pi}_{2m+2}) = \Delta(E_{\mathring{m}},E_{\mathring{m}},E_{2m+2},E_{2m+2},E_{m+1},E_{\mathring{m}},E_{m+1},E_{\mathring{m}}) \quad , \quad \sigma(\bar{\pi}_{2m+2}) = \Delta(E_{2m+2},E_{\mathring{m}},E_{\mathring{m}},E_{\mathring{m}},E_{\mathring{m}}).$$

In view of Lemmata 3.2 and 3.3 it now suffices to prove that, for each $\varepsilon \in E$, the set of objects $\{T(\varepsilon)_n, T(\bar{\varepsilon})_n \mid n \in \mathbb{N}\}$ together with the set of morphisms $\{\iota_n, \bar{\iota}_n, \pi_n, \bar{\pi}_n \mid n \geq 2\}$ satisfies the conditions (c_1) - (c_6) from Lemma 3.2. Indeed, an elementary verification shows that this is true. (As to condition (c_1): Recall from Propositions 1.1 and 1.2 that there are representation equivalences $\phi_0 : {}_\Lambda L \longrightarrow F(K_0)$ and $\phi_1 : \check{F}_0 \longrightarrow \mathcal{M}$ which commute with the Auslander-Reiten translation τ. Moreover, in the lattice category ${}_\Lambda L$ the Auslander-Reiten translation coincides with Heller's operator Ω. Hence, in order to show that $\tau(T(\varepsilon)_1) = T(\bar{\varepsilon})_1$, verify that $\Omega(\phi_0^{-1}\phi_1^{-1}(T(\varepsilon)_1)) = \phi_0^{-1}\phi_1^{-1}(T(\bar{\varepsilon})_1)$.

q.e.d.

LEMMA 3.5. If $S \in \mathrm{ind}\hat{S}$ and $T \in \mathrm{ind}\mathcal{T}(E)$ then S and T are non--isomorphic in $\mathcal{M}$.

PROOF. For the purpose of this proof we need several functors $\Psi_i : \mathcal{M}_i \longrightarrow \mathrm{rep}(k, Q_i)$, $i = 1,2,3$, defined on full subcategories $\mathcal{M}_i$ of $\mathcal{M}$. We introduce them below by prescribing their effect on objects in $\mathcal{M}_i$, thus simultaneously defining $\mathcal{M}_i$ to be the full subcategory of $\mathcal{M}$ which consists of all objects of the indicated type. The functoriality of Ψ_i follows easily from the structure of morphisms in $\mathcal{M}$ together with the structure of objects in $\mathcal{M}_i$. In hatching parts of matrices we follow the convention introduced in the proof of Proposition 2.1. Bars denote the passage from R_3-matrices to k-matrices by reduction modulo π.

$\Psi_1 : \mathcal{M} = \mathcal{M}_1 \longrightarrow \mathrm{rep}(k, Q_1)$

$\Psi_{1'} : \mathcal{M}_{1'} \longrightarrow \mathrm{rep}(k, S_1)$

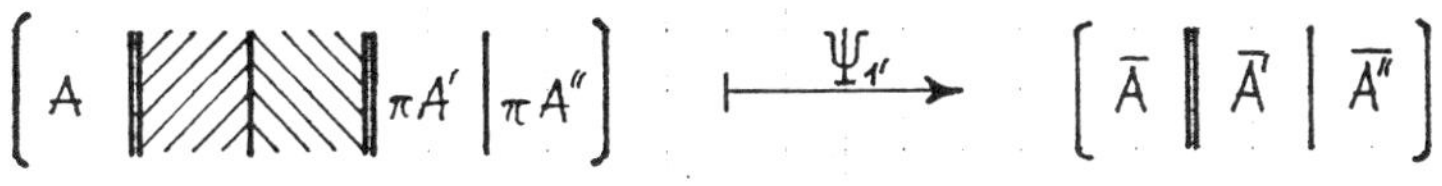

$\Psi_{2,\infty} : \mathcal{M}_{2,\infty} \longrightarrow \mathrm{rep}(k, Q_2)$

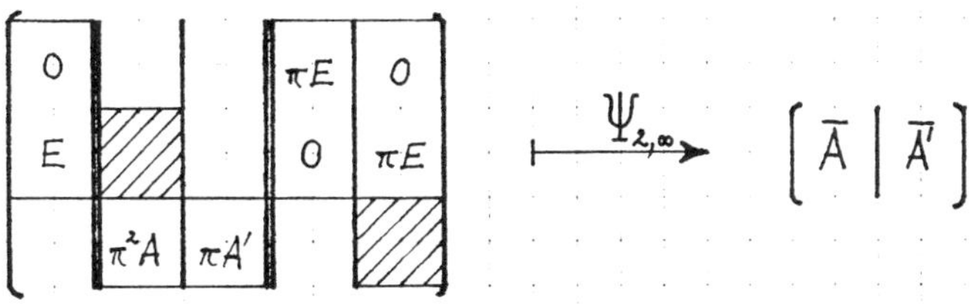

$\Psi_{3,\infty} : \mathcal{M}_{3,\infty} \longrightarrow \mathrm{rep}(k, Q_3)$

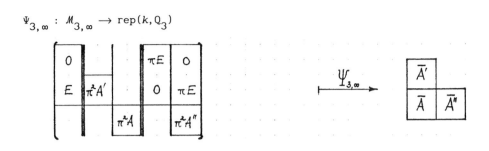

$\Psi_{2,0} : \mathcal{M}_{2,0} \longrightarrow \mathrm{rep}(k, Q_2)$

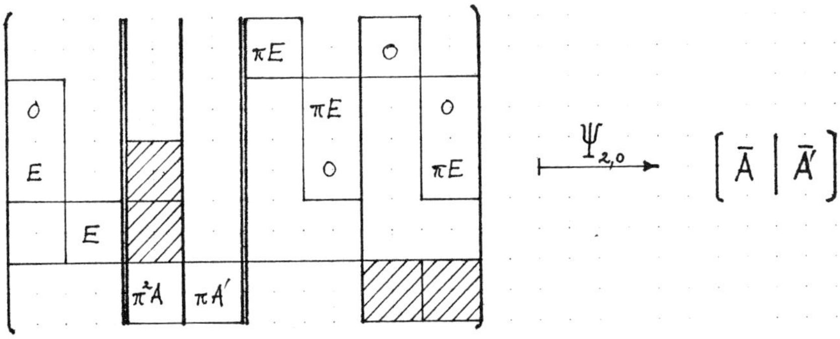

$\Psi_{3,0} : \mathcal{M}_{3,0} \longrightarrow \mathrm{rep}(k, Q_3)$

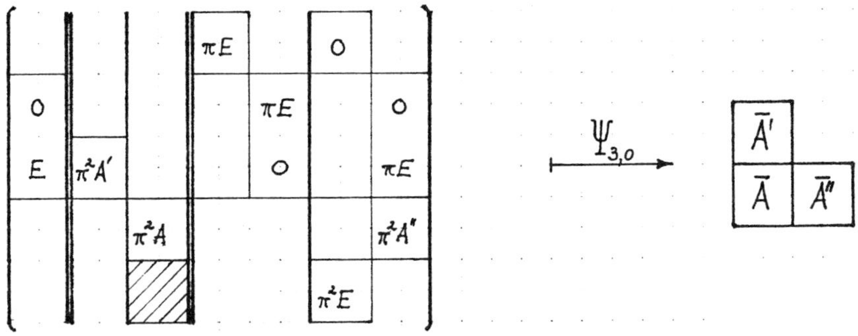

$\Psi_{3,0'} : \mathcal{M}_{3,0'} \longrightarrow \mathrm{rep}(k, Q_3)$

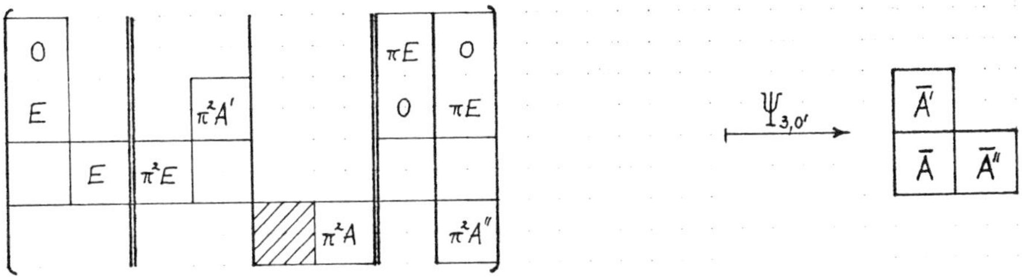

$\Psi_{2,\delta} : \mathcal{M}_{2,\delta} \longrightarrow \mathrm{rep}(k, Q_2)$

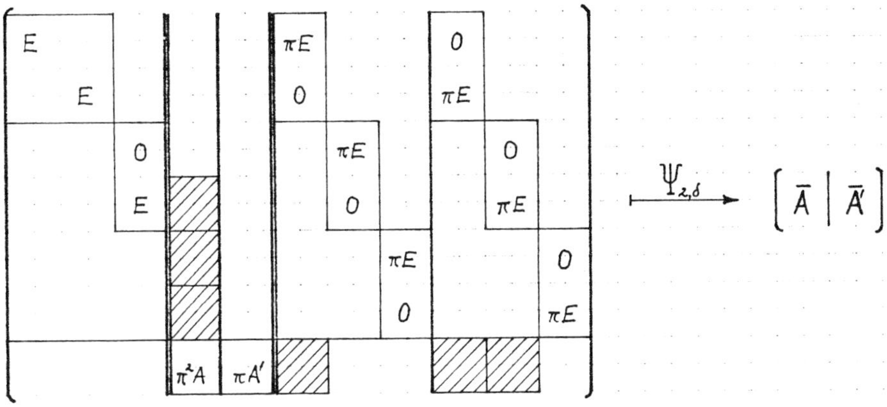

$\Psi_{3,\delta} : \mathcal{M}_{3,\delta} \longrightarrow \mathrm{rep}(k, Q_3)$

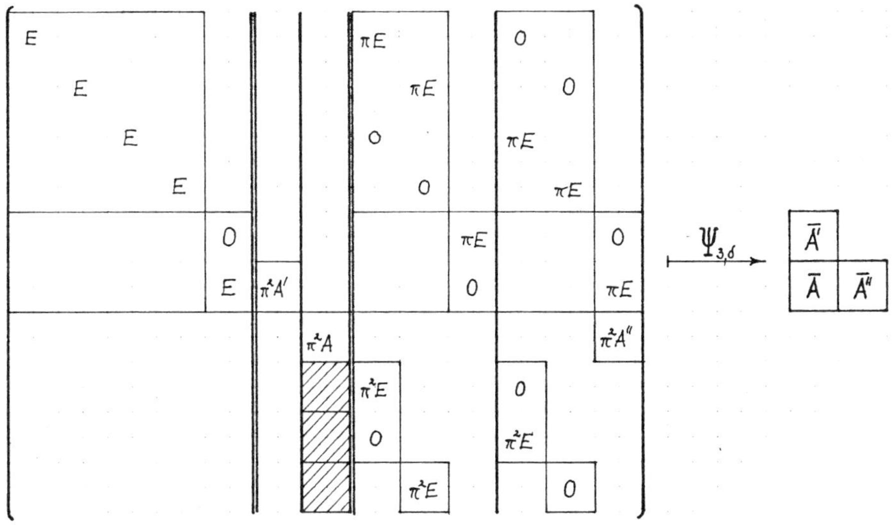

$\Psi_{3,\delta'} : \mathcal{M}_{3,\delta'} \longrightarrow \mathrm{rep}(k, Q_3)$

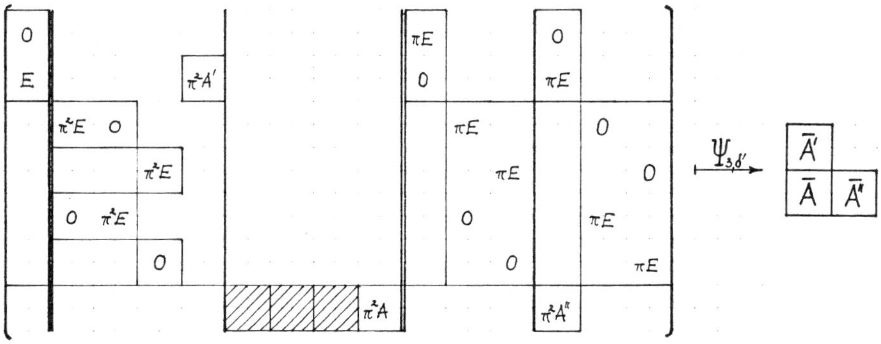

Now let $S \in \mathrm{ind}\hat{S}$, $T \in \mathrm{ind}T(E)$, and assume that $S \cong T$ in $\mathcal{M}$.

Suppose that $T \in \mathrm{ind}T(\infty)$. Then we conclude that $\Psi_1(S) \cong \Psi_1(T) = (0)$, hence $S, T \in \mathcal{M}_{1'}$. Also $\Psi_{1'}(S) \cong \Psi_{1'}(T) = \begin{pmatrix} 0 & E & 0 \\ E & 0 & E \\ 0 & 0 & 0 \end{pmatrix}$, hence $S, T \in \mathcal{M}_{2,\infty}$. Moreover $\Psi_{2,\infty}(S) \cong \Psi_{2,\infty}(T) = (0|0)$, hence $S = \sigma(X,Y,Z)$, where the angular matrix $(X,Y,Z) = \{X_0, X_{0'}, X_5, X_8, X_{8'}, Y_2, Y_4, Y_{4'}\}$ satisfies $X_0 = 0$ and $X_{0'} \equiv 0 \pmod{\pi}$.

In view of Proposition 4.4 we may assume that $X_{0'} = 0$. Therefore $S = S_1 = \sigma\{X_5, X_8, X_{8'}\}$ or $S = S_2 = \sigma\{Y_2, Y_4, Y_{4'}\}$, and $T \cong S_1$ or $T \cong S_2$. All objects S_1, S_2, T are in $\mathcal{M}_{3,\infty}$, hence $\Psi_{3,\infty}(T) \cong \Psi_{3,\infty}(S_1)$ or $\Psi_{3,\infty}(T) \cong \Psi_{3,\infty}(S_2)$. However $\Psi_{3,\infty}(T) \in \mathrm{reg}_{\infty} kQ_3$, by definition of $T(\infty)$. On the other hand $\Psi_{3,\infty}(S_1) = \rho\{X_5, X_8, X_{8'}\}$ and $\Psi_{3,\infty}(S_2) = \lambda\{Y_2, Y_4, Y_{4'}\}$ are not in $\mathrm{reg}_{\infty} kQ_3$, by definition of $\hat{S}$. This is a contradiction.

Suppose that $T \in \mathrm{ind}T(\varepsilon)$, with $\varepsilon = 0, \delta$. Then, by way of analogous arguments, application of the functors $\Psi_1, \Psi_{1'}, \Psi_{2,\varepsilon}, \Psi_{3,\varepsilon}, \Psi_{3,\varepsilon'}$ leads to a contradiction.

Therefore S and T have to be non-isomorphic. q.e.d.

COROLLARY 3.6. (Third reduction). The quivers $A_{\mathcal{M}}(\hat{S} \vee T(E))$ and $A(\hat{S}) \dot{\cup} A_{\mathcal{M}}(T(E))$ are isomorphic. The quiver $A_{\mathcal{M}}(T(E))$ consists of three components, each of which is a stable tube of rank 2.

PROOF. By Proposition 3.4 and Lemma 3.5 we know that $A_{\mathcal{M}}(\hat{S} \vee T(E)) = A_{\mathcal{M}}(\hat{S}) \dot{\cup} A_{\mathcal{M}}(T(E))$ and that $A_{\mathcal{M}}(T(E)) = \dot{\bigcup}_{\varepsilon \in E} A_{\mathcal{M}}(T(\varepsilon))$, where each of the quivers $A_{\mathcal{M}}(T(\varepsilon))$ is a stable tube of rank 2. Hence it remains to show that $A_{\mathcal{M}}(\hat{S}) = A(\hat{S})$.

We claim that the following assertion (*) is true.

(*) For all $M, M' \in \operatorname{ind}\hat{S}$ and $T \in T(E)$, $\mathcal{M}(T,M')\mathcal{M}(M,T) \subset \operatorname{rad}^2_{\hat{S}}(M,M')$.

Indeed, in case M is isomorphic to one of the objects $(K_1, R_3, (0|1))$, $(K_i, 0, 0)$, $i = 0,1,2$ and M' is isomorphic to one of the objects $(K_0, R_3, (1))$, $(0, R_3, 0)$, $\left(K_1, R_3^2, \left(\begin{array}{c|c} \pi^2 & 0 \\ 0 & 1 \end{array}\right)\right)$, $\left(K_2, R_3^2, \left(\begin{array}{c|c} \pi & 0 \\ 0 & \pi \end{array}\right)\right)$, then (*) is easily verified by direct calculation. In the other case there exists either an Auslander-Reiten sequence starting in M or an Auslander-Reiten sequence ending in M', and (*) follows from the factorization property of Auslander-Reiten sequences.

From (*) we conclude that for all $M, M' \in \operatorname{ind}\hat{S}$, $\operatorname{rad}^2_{\mathcal{M}}(M,M') = \operatorname{rad}^2_{\hat{S}}(M,M')$ and hence $\operatorname{irr}_{\mathcal{M}}(M,M') = \operatorname{irr}_{\hat{S}}(M,M')$. This implies $A_{\mathcal{M}}(\hat{S}) = A(\hat{S})$. q.e.d.

4. FOURTH REDUCTION

In this fourth reduction we construct a functor $\phi : \hat{S} \longrightarrow [F(K)]$, where $[F(K)]$ is the quotient of $F(K)$ with respect to a stable system of relations of $F(K)$, and we show that ϕ is a representation equivalence onto the full subcategory $[\hat{C}]$ of $[F(K)]$. As a consequence, the problem of determining $A(\hat{S})$ will be reduced to the problem of determining $A(\hat{C})$. We point out that, although we are still working on the base-dependent level, the approach in this fourth reduction is again entirely different from the approaches in the second and third reduction. Instead of reducing matrices or constructing Auslander-Reiten components we shall construct and analyse the functor ϕ. Eventually this will amount to a very arduous investigation of a system of 435 congruences of quadratic polynomials in matrix blocks.

First of all we have to clarify the definition of $[F(K)]$. Recall from 0.2 the definition of $F(K)$, $F(K_{[-3,-1]})$ and $F(K_{[0,8]})$. We chose k-bases $\{x_i\}$ for all one-dimensional indecomposable objects K_i of K, and $\{x_j, x_{j'}\}$ for all two-dimensional indecomposable objects K_j of K. In the sequel we write $F = F(K)$, $F_u = F(K_{[-3,-1]})$, $F_f = F(K_{[0,8]})$, for brevity. We introduce some more notation and we collect some basic observations centering around the definition of $[F(K)]$.

(i) There are canonical functors $\phi_u : F \longrightarrow F_u$ and $\phi_f : F \longrightarrow F_f$, given as follows. Let $X = (K, F, \varphi)$ and $X' = (K', F', \varphi')$ be objects in F, with $K = \overset{8}{\underset{i=-3}{\oplus}} K_i^{n_i}$ and $K' = \overset{8}{\underset{i=-3}{\oplus}} K_i^{n'_i}$, and let $(\beta, \alpha) : X \longrightarrow X'$ be a morphism in F. Then $\phi_u(X) = (K_u, F_u, \varphi_u) = (\overset{-1}{\underset{i=-3}{\oplus}} K_i^{n_i}, \operatorname{im}(\varphi\iota), \varphi\iota)$ and $\phi_u(\beta, \alpha) = (\beta_u, \alpha_u) = (\beta\iota, \alpha\kappa)$, where

$\iota : K_u \longrightarrow K$ and $\kappa : F_u \longrightarrow F$ are the inclusion mappings. Also, $\phi_f(X) = (K_f, F_f, \varphi_f) = (\bigoplus_{i=0}^{8} K_i^{n_i}, F/F_u, \varphi/\varphi_u)$ and $\phi_f(\beta,\alpha) = (\beta_f, \alpha_f) = (\beta/\beta_u, \alpha/\alpha_u)$. Frequently we shall write $X_u = \phi_u(X)$ and $X_f = \phi_f(X)$.

(ii) Every object X in F determines the short exact sequence $X_u \xrightarrow{(\iota,\kappa)} X \xrightarrow{(\bar\iota,\bar\kappa)} X_f$ in F , where (ι,κ) is defined as in (i) and where $(\bar\iota,\bar\kappa) = (\mathrm{coker}\iota, \mathrm{coker}\kappa)$.

(iii) Every morphism $(\beta,\alpha) : X \longrightarrow X'$ in F determines the following commutative diagram in mod$k$.

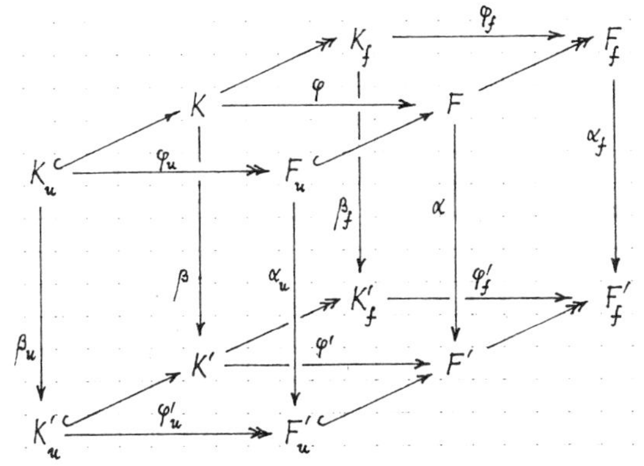

(iv) We define $\mathfrak{R} = \{\mathfrak{R}(X,X') \mid X,X' \in F\}$ to be the system of relations for F which is given by

$\mathfrak{R}(X,X') = \{(\rho,0) \in F(X,X') \mid \mathrm{im}\rho_f \subset kx_{0'}^{n'_0}$ and $\ker\rho_f \supset \bigoplus_{i\in I} K_i^{n_i} \oplus kx_8^{n_8}\}$,

$I = \{0,1,2,3,4,6,7\}$, for all objects $X = (K,F,\varphi)$ and $X' = (K',F',\varphi')$ in F , with $K = \bigoplus_{i=-3}^{8} K_i^{n_i}$ and $K' = \bigoplus_{i=-3}^{8} K_i^{n'_i}$. It is easily verified that $\mathfrak{R}$ is a stable system of relations for F . That is to say, for all $X,X',X'' \in F$ the system of relations $\mathfrak{R}$ satisfies the following conditions:

(r_0) $\mathfrak{R}(X,X')$ is a k-subspace of the k-vectorspace $F(X,X')$.

(r_1) $\mathfrak{R}(X',X'')F(X,X') \subset \mathfrak{R}(X,X'')$.

(r_2) $F(X',X'')\mathfrak{R}(X,X') \subset \mathfrak{R}(X,X'')$.

DEFINITION (The categories $[F(K)]$ and $[\hat{C}]$) . We define $[F(K)]$ to be the quotient category of $F(K)$ with respect to the stable system of relations $\mathfrak{K}$. Moreover, we define $[\hat{C}]$ to be the full subcategory of $[F(K)]$ whose class of objects is the class of objects of $\hat{C}$.

(v) Since $\mathfrak{K}$ is a stable system of relations, the category $[F(K)]$ is well-defined. We write $[F] = [F(K)] = F(K)/\mathfrak{K}$.
(vi) For all $(\beta,\alpha) \in F(X,X')$ we write $[\beta,\alpha] = (\beta,\alpha) + \mathfrak{K}(X,X')$ for the morphism in $[F](X,X')$ which is represented by (β,α) .
(vii) In particular, if (β,α) , $(\beta',\alpha') \in F(X,X')$ are morphisms such that $\beta_f = \beta'_f$ and $\alpha = \alpha'$ then $[\beta,\alpha] = [\beta',\alpha']$. If the converse of this implication would also be true then the definition of $\mathfrak{K}$ would become much more natural. However, our definition of $\mathfrak{K}$ is only rather close to that, and it will become clear during the proof of Proposition 4.4 that this seemingly involved definition of $\mathfrak{K}$ is forced.
(viii) By definition of $\hat{C}$, for which the reader is referred to 0.2, with any object (K,F,φ) in $\hat{C}$ we are given canonical k-bases for K and F . Hence if we consider a diagram in mod$k$ which depends on given objects (K,F,φ) and (K',F',φ') in $\hat{C}$, then all morphisms in the diagram may be identified with matrices, corresponding to them with respect to the given bases of K , K′ , F and F′ . We denote by (μ) the matrix which in this sense corresponds to a k-linear morphism μ .
(ix) Furthermore, if (K,F,φ) and (K',F',φ') are objects in $\hat{C}$ then, identifying F_f and F'_f with the canonical direct summands of F and F′ complementary to F_u and F'_u , we have the decompositions $K = K_u \oplus K_f$, $K' = K'_u \oplus K'_f$ and $F = F_u \oplus F_f$, $F' = F'_u \oplus F'_f$. We write

$\varphi = \begin{pmatrix} \varphi_u \varphi_e \\ 0 \ \varphi_f \end{pmatrix}$, $\varphi' = \begin{pmatrix} \varphi'_u \varphi'_e \\ 0 \ \varphi'_f \end{pmatrix}$, and $\beta = \begin{pmatrix} \beta_u \beta_e \\ 0 \ \beta_f \end{pmatrix}$ for any morphism $\beta \in K(\mathrm{K},\mathrm{K}')$, and $\alpha = \begin{pmatrix} \alpha_u \alpha_e \\ 0 \ \alpha_f \end{pmatrix}$ for any morphism $\alpha \in \mathrm{Hom}_k(\mathrm{F},\mathrm{F}')$, referring to these decompositions of $\mathrm{K},\mathrm{K}',\mathrm{F}$ and F' .

The next two Lemmata together with the ensuing Corollary are technical preparations for the construction of the functor $\phi : \hat{S} \longrightarrow [\hat{C}]$ which we are aiming at.

LEMMA 4.1. Let $X = (\mathrm{K},\mathrm{F},\varphi)$ and $X' = (\mathrm{K}',\mathrm{F}',\varphi')$ be objects in $\hat{C}$, together with their associated objects $X_u = (\mathrm{K}_u,\mathrm{F}_u,\varphi_u)$ and $X'_u = (\mathrm{K}'_u,\mathrm{F}'_u,\varphi'_u)$ in F_u . Then we have:

(i) Morphisms $\alpha_u \in \mathrm{Hom}_k(\mathrm{F}_u,\mathrm{F}'_u)$ and $\beta_u \in K_u(\mathrm{K}_u,\mathrm{K}'_u)$ correspond to k-matrices (α_u) and (β_u) having block structure given by

$$(\alpha_u) = \begin{bmatrix} A_{11} & \cdots & A_{17} \\ \vdots & & \vdots \\ A_{71} & \cdots & A_{77} \end{bmatrix} \quad , \quad (\beta_u) = \left[\begin{array}{cc|cc|cc|cc} B_{11} & B_{12} & & & & & & \\ B_{21} & B_{22} & & & & & & \\ \hline & & B_{33} & B_{34} & & & & \\ & & B_{43} & B_{44} & & & & \\ \hline & & & & B_{55} & B_{56} & B_{57} & B_{58} \\ & & & & B_{65} & B_{66} & B_{67} & B_{68} \\ \hline & & & & dB_{57} & dB_{58} & B_{55} & B_{56} \\ & & & & dB_{67} & dB_{68} & B_{65} & B_{66} \end{array}\right]$$

(ii) Let $\alpha_u \in \mathrm{Hom}_k(\mathrm{F}_u,\mathrm{F}'_u)$ and $\beta_u \in K_u(\mathrm{K}_u,\mathrm{K}'_u)$. Then $(\beta_u,\alpha_u) \in F_u(X_u,X'_u)$ if and only if the corresponding k-matrices (α_u) and (β_u) satisfy the equations

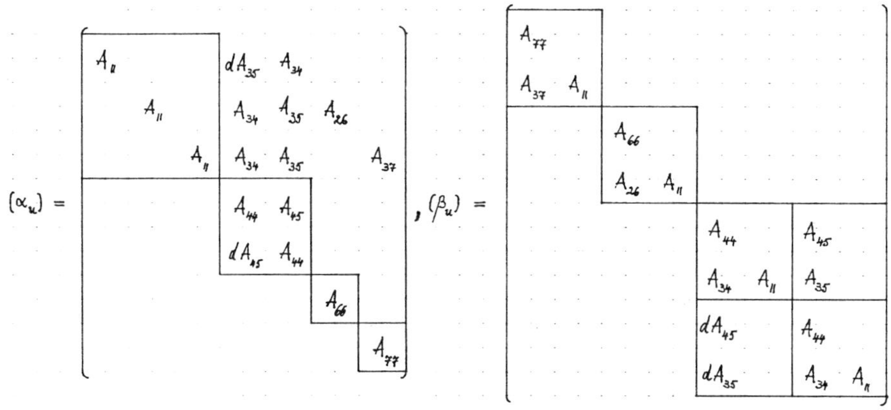

(iii) Let $\alpha_u \in \mathrm{Hom}_k(F_u, F'_u)$ be given by a k-matrix (α_u) of the special block structure described in (ii). Then there exists a uniquely determined morphism $\beta_u \in K_u(K_u, K'_u)$ such that $(\beta_u, \alpha_u) \in F_u(X_u, X'_u)$.

(iv) Let $(\beta, \alpha) \in F(X, X')$. Then the k-matrices (α) and (β_f) corresponding to α and β_f have the block structure given by

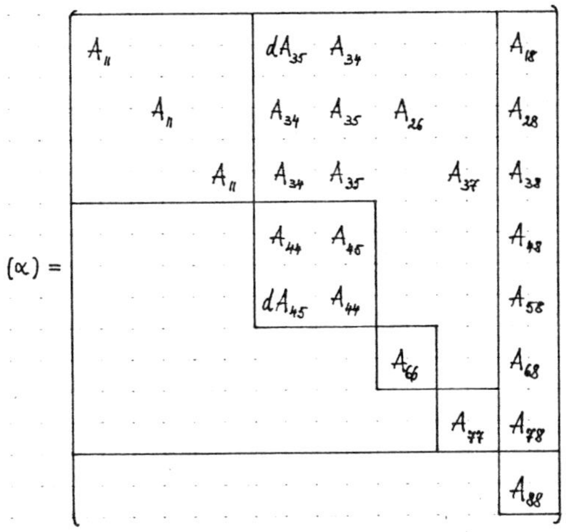

$$
(\beta_f) = \left(\begin{array}{cc|c|c|cc|cc|c|c|cc|cc}
B_{99} & & B_{9,11} & & B_{9,13} & B_{9,14} & B_{9,15} & B_{9,16} & B_{9,17} & B_{9,18} & B_{9,19} & B_{9,20} & B_{9,21} & B_{9,22} \\
 & B_{99} & & B_{10,12} & dB_{9,14} & B_{9,13} & B_{10,15} & B_{9,15} & B_{10,17} & B_{10,18} & B_{10,19} & B_{10,20} & B_{10,21} & B_{10,22} \\
\hline
 & & B_{11,11} & & & & & B_{11,16} & & B_{11,18} & B_{11,19} & B_{11,20} & B_{11,21} & B_{11,22} \\
\hline
 & & & B_{12,12} & & & B_{12,15} & & B_{12,17} & & B_{12,19} & B_{12,20} & B_{12,21} & B_{12,22} \\
\hline
 & & & & B_{13,13} & B_{13,14} & B_{13,15} & B_{13,16} & B_{13,17} & B_{13,18} & B_{13,19} & B_{13,20} & B_{13,21} & B_{13,22} \\
 & & & & dB_{13,14} & B_{13,13} & dB_{13,16} & B_{13,15} & B_{14,17} & B_{14,18} & dB_{13,20} & B_{13,19} & B_{14,21} & B_{14,22} \\
\hline
 & & & & & & B_{15,15} & & B_{15,17} & & B_{15,19} & B_{15,20} & B_{15,21} & B_{15,22} \\
 & & & & & & & B_{15,15} & & B_{16,18} & dB_{15,20} & B_{15,19} & B_{16,21} & B_{15,21} \\
\hline
 & & & & & & & & B_{17,17} & & & & & B_{17,22} \\
\hline
 & & & & & & & & & B_{18,18} & & & B_{18,21} & \\
\hline
 & & & & & & & & & & B_{19,19} & B_{19,20} & B_{19,21} & B_{19,22} \\
 & & & & & & & & & & dB_{19,20} & B_{19,19} & dB_{19,22} & B_{19,21} \\
\hline
 & & & & & & & & & & & & B_{21,21} & \\
 & & & & & & & & & & & & & B_{21,21}
\end{array}\right)
$$

PROOF. (i) By assumption, X and X' are objects in $\hat{C}$. Hence, by definition of $\hat{C}$, (φ_u) and (φ'_u) have prescribed normal form. The block partitions (α_u) and (β_u) correspond to the block partitions of (φ_u) and (φ'_u).

(ii) We know that $(\beta_u, \alpha_u) \in F_u(X_u, X'_u)$ if only if $(\varphi'_u)(\beta_u) = (\alpha_u)(\varphi_u)$, where (α_u) and (β_u) are of the form described in (i). Multiplication of matrices shows that this equation is true if and only if (α_u) and (β_u) satisfy the conditions formulated in (ii).

(iii) This is an immediate consequence of (ii).

(iv) The asserted form of (α) follows from $(\alpha) = \begin{pmatrix} \alpha_u & \alpha_e \\ 0 & \alpha_f \end{pmatrix}$, together with (ii). The asserted form of (β_f) follows from the fact that $\beta_f \in K_f(K_f, K_f')$, together with the definition of K_f. q.e.d.

LEMMA 4.2. Consider the commutative diagram

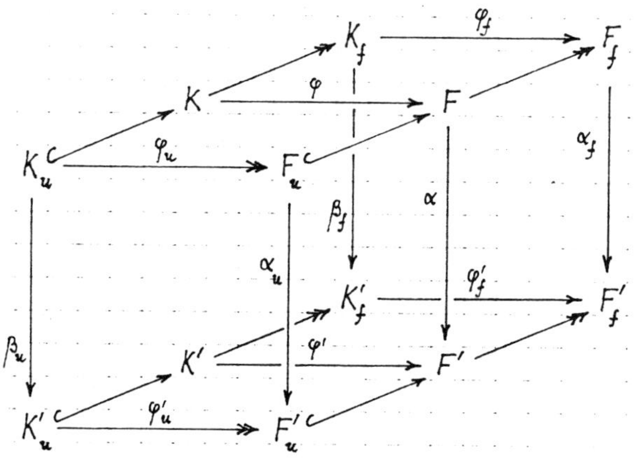

where $X = (K, F, \varphi)$, $X' = (K', F', \varphi') \in \hat{C}$ and $(\beta_u, \alpha_u) \in F_u(X_u, X_u')$, $(\beta_f, \alpha_f) \in F_f(X_f, X_f')$. Then we have:

(i) If there exists a morphism $\beta \in K(K, K')$ which completes the diagram fully commutatively, then $[\beta, \alpha] \in [\hat{C}](X, X')$ is uniquely determined.

(ii) There exists a morphism $\beta \in K(K, K')$ which completes the diagram fully commutatively if and only if the k-matrices (α), (β_f), (φ), (φ') satisfy the following conditions (recall that by assumption $(\varphi) = \mu_\zeta(X, Y, Z)$ and $(\varphi') = \mu_\zeta(X', Y', Z')$, with (X, Y, Z), $(X', Y', Z') \in \mathfrak{M}$) :

$$
(b_0)\qquad
\left(\begin{array}{cc|cc|c|c|cc}
A_{11} & & -d^{-1}A_{34} & -A_{35} & A_{26} & & -d^{-1}A_{18} & A_{28} \\
 & A_{11} & A_{35} & A_{34} & & A_{37} & A_{38} & -A_{48} \\
\hline
 & & A_{44} & dA_{45} & & & A_{58} & -dA_{48} \\
 & & A_{45} & A_{44} & & & A_{48} & -A_{58} \\
\cline{3-8}
 & & & & A_{66} & & & A_{68} \\
\cline{5-8}
 & & & & & A_{77} & A_{78} & \\
\cline{6-8}
 & & & & & & A_{88} & \\
 & & & & & & & A_{88}
\end{array}\right)
\begin{pmatrix} Y_4 \\ Y_{4'} \\ Y_3 \\ Y_{3'} \\ Y_2 \\ Y_1 \\ X_0 \\ X_{0'} \end{pmatrix}
=
\begin{pmatrix} Y'_4 \\ Y'_{4'} \\ Y'_3 \\ Y'_{3'} \\ Y'_2 \\ Y'_1 \\ X'_0 \\ X'_{0'} \end{pmatrix}
\begin{bmatrix} B_{99} \end{bmatrix} .
$$

$$(b_1)\qquad A_{77}Z_1 + A_{78}X_1 = Z'_1B_{11,11} + Y'_1B_{9,11} .$$

$$(b_2)\qquad A_{66}Z_2 + A_{68}X_2 = Z'_2B_{12,12} + Y'_2B_{10,12} .$$

$$(b_3)\qquad \begin{pmatrix} A_{44} & A_{45} \\ dA_{45} & A_{44} \end{pmatrix}\begin{pmatrix} Z_{3'3} & Z_{3'3'} \\ Z_{33} & Z_{33'} \end{pmatrix} + \begin{pmatrix} A_{48} \\ A_{58} \end{pmatrix}(X_3X_{3'}) =$$

$$= \begin{pmatrix} Z'_{3'3} & Z'_{3'3'} \\ Z'_{33} & Z'_{33'} \end{pmatrix}\begin{pmatrix} B_{13,13} & B_{13,14} \\ dB_{13,14} & B_{13,13} \end{pmatrix} + \begin{pmatrix} Y'_{3'} \\ Y'_3 \end{pmatrix}(B_{9,13}B_{9,14}) +$$

$$+ \begin{pmatrix} R & S \\ dS & R \end{pmatrix} ,\ \text{for some } k\text{-matrices } R \text{ and } S .$$

PROOF. (i) This follows immediately from remark (vii) above.

(ii) There exists a morphism $\beta \in K(K,K')$ which completes the diagram fully commutatively if and only if there exists a morphism $\beta_e \in K(K_f,K'_u)$ such that $\beta = \begin{pmatrix} \beta_u & \beta_e \\ 0 & \beta_f \end{pmatrix}$ satisfies the equation $\alpha\varphi = \varphi'\beta$. In turn, this is the case if and only if there exists a morphism $\beta_e \in K(K_f,K'_u)$ such that (β_e) satisfies the equation

$$(*)\quad (\varphi'_u)(\beta_e) = (\alpha_u)(\varphi_e) + (\alpha_e)(\varphi_f) - (\varphi'_e)(\beta_f) ,\ \text{where}\ (\alpha) = \begin{pmatrix} \alpha_u & \alpha_e \\ 0 & \alpha_f \end{pmatrix} .$$

Here, the matrices (φ_e), (φ_f) (φ'_u), (φ'_e) have the prescribed forms which are inherent in $(\varphi) = \mu_\zeta(X,Y,Z)$ and $(\varphi') = \mu_\zeta(X',Y',Z')$, and the matrices (α) and (β_f) may be assumed to be of the form described in Lemma 4.1, (iv). Hence, on setting $(\beta_e) = (B_{ij})_{\substack{i=1,\dots,7 \\ j=9,\dots,22}}$, equation (*) amounts to a system of 98 linear matrix-equations in the indeterminates B_{ij}. Calculation shows that the system of linear equations (*) is solvable in B_{ij} if and only if the following sub-system (*′) is solvable in B_{ij} :

$$dA_{35}Y_{3'} + A_{34}Y_3 + A_{18}X_0 = dB_{6,10}$$

$$A_{11}Y_4 + A_{26}Y_2 + A_{28}X_{0'} - Y'_4B_{99} = B_{6,10}$$

$$A_{18}X_{0'} = B_{69}$$

$$A_{11}Y_{4'} + A_{34}Y_{3'} + A_{35}Y_3 + A_{37}Y_1 + A_{38}X_0 - Y'_{4'}B_{99} = B_{69}$$

$$A_{48}X_{0'} = B_{5,10}$$

$$dA_{45}Y_{3'} + A_{44}Y_3 + A_{58}X_0 - Y'_3B_{99} = dB_{5,10}$$

$$A_{58}X_{0'} = B_{59}$$

$$A_{44}Y_{3'} + A_{45}Y_3 + A_{48}X_0 - Y'_{3'}B_{99} = B_{59}$$

$$A_{66}Y_2 + A_{68}X_{0'} - Y'_2B_{99} = 0$$

$$A_{77}Y_1 + A_{78}X_0 - Y'_1B_{99} = 0$$

$$A_{77}Z_1 + A_{78}X_1 - (Y'_1B_{9,11} + Z'_1B_{11,11}) = 0$$

$$A_{66}Z_2 + A_{68}X_2 - (Y'_2B_{10,12} + Z'_2B_{12,12}) = 0$$

$$A_{44}Z_{3'3} + A_{45}Z_{33} + Z_{48}X_3 - (Y'_{3'}B_{9,13} + Z'_{3'3}B_{13,13} + Z'_{3'3'}dB_{13,14}) = B_{5,13}$$

$$dA_{45}Z_{3'3'} + A_{44}Z_{33'} + A_{58}X_{3'} - (Y'_3B_{9,14} + Z'_{33}B_{13,14} + Z'_{33'}B_{13,13}) = B_{5,13}$$

$$A_{44}Z_{3'3'} + A_{45}Z_{33'} + A_{48}X_{3'} - (Y'_{3'}B_{9,14} + Z'_{3'3}B_{13,14} + Z'_{3'3'}B_{13,13}) = B_{5,14}$$

$$dA_{45}Z_{3'3} + A_{44}Z_{33} + A_{58}X_3 - (Y'_3B_{9,13} + Z'_{33}B_{13,13} + Z'_{33'}dB_{13,14}) = dB_{5,14}$$

Now it is trivial to verify that the system of linear equations (*′) is solvable in B_{ij} if and only if the matrices (α), (β_f) (φ), (φ') satisfy the conditions (b_0) - (b_3). q.e.d.

COROLLARY 4.3. Let $X = (K,F,\varphi)$ and $X' = (K',F',\varphi')$ be objects in $\hat{C}$. Let $((\beta_f),(\alpha))$ be a pair of k-matrices satisfying the following conditions:

(m_0) (α) and (β_f) are of the form described in Lemma 4.1, (iv).

(m_1) $(\alpha_f)(\varphi_f) = (\varphi'_f)(\beta_f)$.

(m_2) (α), (β_f), (φ), (φ') satisfy conditions (b_0)-(b_3) from Lemma 4.2, (ii).

Then there exists a uniquely determined morphism $\left[\begin{pmatrix} \beta_u & \beta_e \\ 0 & \beta_f \end{pmatrix}, \alpha\right] \in [\hat{C}](X,X')$.

PROOF. Due to (m_0) and Lemma 4.1, (iii), there exists a uniquely determined morphism $(\beta_u,\alpha_u) \in F_u(X_u,X'_u)$. Due to (m_0) and (m_1) we know that $(\beta_f,\alpha_f) \in F_f(X_f,X'_f)$. Hence we are in the situation of Lemma 4.2. Due to (m_2) and Lemma 4.2, (ii), there exists a morphism $\left(\begin{pmatrix} \beta_u\beta_e \\ 0\ \beta_f \end{pmatrix}, \alpha\right) \in \hat{C}(X,X')$. Due to Lemma 4.2, (i), the corresponding residue class morphism $\left[\begin{pmatrix} \beta_u\beta_e \\ 0\ \beta_f \end{pmatrix}, \alpha\right]$ in $[\hat{C}](X,X')$ is uniquely determined. q.e.d.

PROPOSITION 4.4. There exists a representation equivalence $\phi : \hat{S} \longrightarrow [\hat{C}]$.

PROOF. We subdivide the proof into four parts.

1) <u>Definition of the functor</u> $\phi : \hat{S} \longrightarrow [\hat{C}]$.

We define ϕ on objects by $\phi(\sigma(X,Y,Z)) = \zeta(X,Y,Z)$, for all $\sigma(X,Y,Z) \in \hat{S}$. (This assignment is well-defined because $\zeta(X,Y,Z)$ depends only on the residue class matrix modulo π of $(X,Y,Z) \in \mathfrak{M}$.)

Let $M = \sigma(X,Y,Z)$ and $M' = \sigma(X',Y',Z')$ be objects in $\hat{S}$, and put $\phi(M) = (K,F,\varphi)$ and $\phi(M') = (K',F',\varphi')$. For definition of ϕ on morphisms our strategy is to assign to any given morphism $(T,S) \in \hat{S}(M,M')$ a pair of k-matrices $((\beta_f),(\alpha))$ which satisfies the conditions (m_0)-(m_2) of Corollary 4.3 and which therefore uniquely determines a morphism $[\beta,\alpha] \in [\hat{C}](\phi(M),\phi(M'))$. Then we put $\phi(T,S) := [\beta,\alpha]$. In order to proceed along this line we need a very careful analysis of morphisms in $\hat{S}$.

Let $(T,S) \in \hat{S}(M,M')$. Then S and T have block structures in accordance with the block structures of M and M',

$$S = \begin{bmatrix} S_{11} & \cdots & S_{1,15} \\ \vdots & & \vdots \\ S_{15,1} & \cdots & S_{15,15} \end{bmatrix}$$

$$T = \left(\begin{array}{ccc||ccc|ccc||ccc|ccc}
T_{11} & \cdots & T_{17} & & & & \pi T_{1,12} & \cdots & \pi T_{1,15} & \pi^2 T_{1,16} & \cdots & \pi^2 T_{1,22} & \pi^2 T_{1,23} & \cdots & \pi^2 T_{1,29} \\
\vdots & & \vdots & & 0 & & \vdots & & \vdots & \vdots & & \vdots & \vdots & & \vdots \\
T_{71} & \cdots & T_{77} & & & & \pi T_{7,12} & \cdots & \pi T_{7,15} & \pi^2 T_{7,16} & \cdots & \pi^2 T_{7,22} & \pi^2 T_{7,23} & \cdots & \pi^2 T_{7,29} \\
\hline\hline
T_{81} & \cdots & T_{87} & T_{88} & \cdots & T_{8,11} & T_{8,12} & \cdots & T_{8,15} & T_{8,16} & \cdots & T_{8,22} & T_{8,23} & \cdots & T_{8,29} \\
\vdots & & \vdots & \vdots & & \vdots & \vdots & & \vdots & \vdots & & \vdots & \vdots & & \vdots \\
T_{11,1} & \cdots & T_{11,7} & T_{11,8} & \cdots & T_{11,11} & T_{11,12} & \cdots & T_{11,15} & T_{11,16} & \cdots & T_{11,22} & T_{11,23} & \cdots & T_{11,29} \\
\hline
T_{12,1} & \cdots & T_{12,7} & \pi^2 d T_{8,12} & \cdots & \pi^2 d T_{8,15} & T_{88} & \cdots & T_{8,11} & \pi d T_{8,23} & \cdots & \pi d T_{8,29} & \pi T_{8,16} & \cdots & \pi T_{8,22} \\
\vdots & & \vdots & \vdots & & \vdots & \vdots & & \vdots & \vdots & & \vdots & \vdots & & \vdots \\
T_{15,1} & \cdots & T_{15,7} & \pi^2 d T_{11,12} & \cdots & \pi^2 d T_{11,15} & T_{11,8} & \cdots & T_{11,11} & \pi d T_{11,23} & \cdots & \pi d T_{11,29} & \pi T_{11,16} & \cdots & \pi T_{11,22} \\
\hline\hline
T_{16,1} & \cdots & T_{16,7} & \pi T_{16,8} & \cdots & \pi T_{16,11} & T_{16,12} & \cdots & T_{16,15} & T_{16,16} & \cdots & T_{16,22} & T_{16,23} & \cdots & T_{16,29} \\
\vdots & & \vdots & \vdots & & \vdots & \vdots & & \vdots & \vdots & & \vdots & \vdots & & \vdots \\
T_{22,1} & \cdots & T_{22,7} & \pi T_{22,8} & \cdots & \pi T_{22,11} & T_{22,12} & \cdots & T_{22,15} & T_{22,16} & \cdots & T_{22,22} & T_{22,23} & \cdots & T_{22,29} \\
\hline
T_{23,1} & \cdots & T_{23,7} & \pi d T_{16,12} & \cdots & \pi d T_{16,15} & T_{16,8} & \cdots & T_{16,11} & d T_{16,23} & \cdots & d T_{16,29} & T_{16,16} & \cdots & T_{16,22} \\
\vdots & & \vdots & \vdots & & \vdots & \vdots & & \vdots & \vdots & & \vdots & \vdots & & \vdots \\
T_{29,1} & \cdots & T_{29,7} & \pi d T_{22,12} & \cdots & \pi d T_{22,15} & T_{22,8} & \cdots & T_{22,11} & d T_{22,23} & \cdots & d T_{22,29} & T_{22,16} & \cdots & T_{22,22}
\end{array}\right)$$

and $SM = M'T$. After matrix-multiplication, this equation amounts to a system (*) of 435 congruences modulo π^{ν} $(1 \le \nu \le 3)$ of R_3-matrices, each of them being a quadratic polynomial in the set of blocks of S, X, Y, Z, X', Y', Z' and T . Our task now is to investigate closely the system of congruences (*). As a first consequence we claim:

(1) The matrices $S_{13,15}$, $S_{14,15}$, S_{98} , $T_{9,10}$, $T_{9,11}$, $T_{21,20}$ are congruent 0 modulo π .

In order to prove assertion (1) observe that the following system of congruences modulo π may be deduced from (*):

(1.1) $Y'_j T_{9,10} \equiv 0$ and $Y'_j T_{9,11} \equiv 0$, for all $j \in \check{I}_Y = I_Y \setminus \{2,4,4'\}$.

(1.2) $S_{13,15} X_i \equiv 0$ and $S_{14,15} X_i \equiv 0$, for all $i \in \check{I}_X = I_X \setminus \{5,8,8'\}$.

(1.3) $$\begin{cases} Y'_{4'} T_{9,10} \equiv S_{98} & \text{and } Y'_{4'} T_{9,11} \equiv 0 . \\ S_{13,15} X_8 \equiv T_{21,20} & \text{and } S_{14,15} X_8 \equiv 0 \text{ and } T_{21,20} \equiv S_{98} . \\ S_{13,15} X_{8'} \equiv Y'_4 T_{9,10} & \text{and } S_{13,15} X_5 \equiv Y'_4 T_{9,11} . \\ S_{14,15} X_{8'} \equiv Y'_2 T_{9,10} & \text{and } S_{14,15} X_5 \equiv Y'_2 T_{9,11} . \end{cases}$$

Denote by η'_j, ξ_i, σ_{ij}, τ_{kl} the k-linear maps arising from Y'_j, X_i, S_{ij}, T_{kl} after reduction modulo π . Consider the exact sequences $\ker\eta' \xrightarrow{\kappa'} k^{n_0} \xrightarrow{\eta'} \bigoplus_{j \in \check{I}_Y} k^{m'_j}$ and $\bigoplus_{i \in \check{I}_X} k^{n_i} \xrightarrow{\xi} k^{m_0} \xrightarrow{\zeta} \operatorname{coker}\xi$, where $\eta' = (\eta'_j)_{j \in \check{I}_Y}$ and $\xi = (\xi_i)_{i \in \check{I}_X}$. In these terms we conclude that $\eta'(\tau_{9,10} \tau_{9,11}) = 0$ (due to (1.1)) and $\begin{pmatrix} \sigma_{13,15} \\ \sigma_{14,15} \end{pmatrix} \xi = 0$ (due to (1.2)). Hence $(\tau_{9,10} \tau_{9,11})$ and $\begin{pmatrix} \sigma_{13,15} \\ \sigma_{14,15} \end{pmatrix}$ admit factorizations

$$(\tau_{9,10} \tau_{9,11}) = \kappa'(\tau'_{9,10} \tau'_{9,11}) \text{ and } \begin{pmatrix} \sigma_{13,15} \\ \sigma_{14,15} \end{pmatrix} = \begin{pmatrix} \bar{\sigma}_{13,15} \\ \bar{\sigma}_{14,15} \end{pmatrix} \zeta .$$ Together

with (1.3) we obtain that in the following diagram all squares commute.

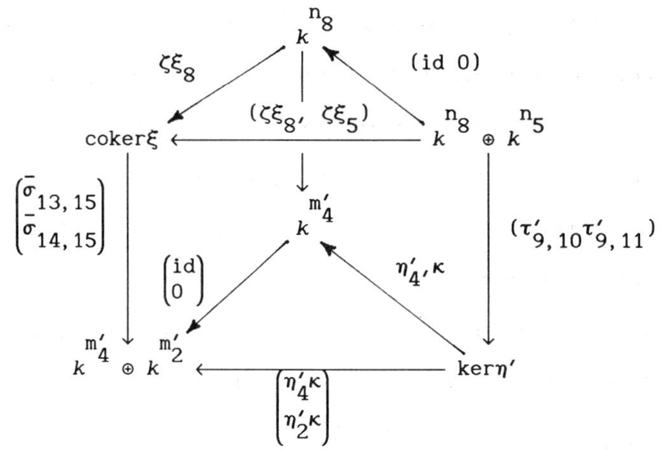

We may rephrase this assertion by saying that $\mu = \left(\begin{pmatrix}\bar{\sigma}_{13,15}\\ \bar{\sigma}_{14,15}\end{pmatrix}, \sigma_{98}, (\tau'_{9,10}\,\tau'_{9,11})\right) : \rho(X,Y,Z) \longrightarrow \lambda(X',Y',Z')$ is a morphism in $\operatorname{rep}(k,Q_3)$. On the other hand, since M and M' are objects in $\hat{S}$, $\lambda(X',Y',Z')$ is preprojective, whereas $\rho(X,Y,Z)$ does not contain any preprojective direct summand. Hence μ has to be zero. This proves (1). (Note that our last argument heavily depends on the definition of $\hat{S}$ and $\hat{\mathfrak{M}}$ which, when given in 0.4, may have seemed to be rather artificial and involved.)

For all $(i,j) = (13,15),(14,15)$ and $(k,l) = (9,10),(9,11)$ let S^*_{ij} and T^*_{kl} be R_3-matrices which satisfy the conditions $S_{ij} = \pi S^*_{ij}$ and $T_{kl} = \pi T^*_{kl}$. By assertion (1) they exist, and they are uniquely determined modulo π. Hence to the given morphism $(T,S) \in \hat{S}(M,M')$ we may assign the following pair $((\beta_f),(\alpha))$ of k-matrices:

$$
(\alpha) \quad = \quad \left(\begin{array}{ccc|cc|c|c|c}
S_{99} & & & dS_{7,10} & dS_{7,11} & & & dS_{7,15} \\
& S_{99} & & dS_{7,11} & S_{7,10} & S_{13,14} & & S^{*}_{13,15} \\
& & S_{99} & dS_{7,11} & S_{7,10} & & S_{9,12} & S_{9,15} \\
\hline
& & & S_{10,10} & S_{10,11} & & & S_{10,15} \\
& & & dS_{10,11} & S_{10,10} & & & S_{11,15} \\
& & & & & S_{14,14} & & S^{*}_{14,15} \\
& & & & & & S_{12,12} & S_{12,15} \\
\hline
& & & & & & & S_{15,15}
\end{array}\right)
$$

$$
(\beta_f) \quad = \quad \left(\begin{array}{cc|c|c|cc|cc|c|c|cc|cc}
T_{99} & & T_{98} & & T_{9,16} & T_{9,23} & T_{9,17} & T_{9,24} & T_{9,15} & T_{9,25} & T_{9,19} & T_{9,26} & T_{9,27} & T_{9,14} \\
& T_{99} & & T_{13,7} & dT_{9,23} & T_{9,16} & dT_{9,24}-T_{13,1} & T_{9,17} & T^{*}_{9,11} & T_{9,18} & dT_{9,26}-T_{13,2} & T_{9,19}-T_{13,3} & T_{9,20}-T_{13,4} & T^{*}_{9,10} \\
\hline
& & T_{88} & & & & & T_{8,24} & & T_{8,25} & T_{8,19} & T_{8,26} & T_{8,27} & T_{8,14} \\
\hline
& & & T_{77} & & & -T_{71} & & T_{7,15} & & -T_{72} & -T_{73} & -T_{74} & T_{7,14} \\
\hline
& & & & T_{16,16} & T_{16,23} & T_{16,17} & T_{16,24} & T_{16,15} & T_{16,25} & T_{16,19} & T_{16,26} & T_{16,27} & T_{16,14} \\
& & & & dT_{16,23} & T_{16,16} & dT_{16,24} & T_{16,17} & T_{16,11} & T_{16,18} & dT_{16,26} & T_{16,19} & T_{16,20} & T_{16,10} \\
\hline
& & & & & & T_{17,17} & & T_{17,15} & & T_{17,19} & T_{17,26} & T_{17,27} & T_{17,14} \\
& & & & & & & T_{17,17} & & T_{17,18} & dT_{17,26} & T_{17,19} & T_{17,20} & T_{17,27} \\
\hline
& & & & & & & & T_{11,11} & & & & & T_{11,10} \\
\hline
& & & & & & & & & T_{18,18} & & & T_{18,20} & \\
\hline
& & & & & & & & & & T_{19,19} & T_{19,26} & T_{19,27} & T_{19,14} \\
& & & & & & & & & & dT_{19,26} & T_{19,19} & dT_{19,14} & T_{19,27} \\
\hline
& & & & & & & & & & & & T_{20,20} & \\
& & & & & & & & & & & & & T_{20,20}
\end{array}\right)
$$

(Clearly, all submatrices S_{ij}, S^*_{ij} of (α) and all submatrices T_{kl}, T^*_{kl} of (β_f) have to be understood modulo π .)

Now the pair $((\beta_f),(\alpha))$ in fact does satisfy the conditions (m_0)-(m_2) formulated in Corollary 4.3: (m_0) is true by definition of $((\beta_f),(\alpha))$, whereas (m_1) and (m_2) can be deduced from the system of congruences (*) (in a process of verification which is long and cumbersome, but trivial in each single step). Hence, on setting $\phi(T,S) = [\beta,\alpha] = \left[\begin{pmatrix}\beta_u & \beta_e \\ 0 & \beta_f\end{pmatrix}, \alpha\right]$, ϕ is well-defined on morphisms.

Let us investigate the behaviour of ϕ with respect to composition of morphisms. For any three objects M, M', M'' in $\hat{S}$ let $(T,S) \in \hat{S}(M,M')$, $(T',S') \in \hat{S}(M',M'')$ and put $\phi(T,S) = [\beta,\alpha]$, $\phi(T',S') = [\beta',\alpha']$, $\phi(T'T,S'S) = [\hat{\beta},\hat{\alpha}]$. Then, by definition of the stable system $\mathfrak{K}$, $\phi((T',S')(T,S)) = \phi(T',S')\phi(T,S)$ if and only if $\hat{\alpha} = \alpha'\alpha$, $\mathrm{im}(\hat{\beta}_f - \beta'_f\beta_f) \subset kx_{0'}^{n''_0}$, and $\ker(\hat{\beta}_f - \beta'_f\beta_f) \supset \bigoplus_{i\in I} K_i^{n_i} \oplus kx_8^{n_8}$. However, the morphisms $(\hat{\beta},\hat{\alpha})$ and $(\beta'\beta,\alpha'\alpha)$ in $\hat{C}(\phi(M),\phi(M''))$ in fact do satisfy this condition, and therefore ϕ respects composition. (For verification one has to use the definition of ϕ on morphisms and the distribution of blocks in S and T , S' and T' , which are congruent 0 modulo π (respectively modulo π^2) , where this latter information again has to be deduced from the congruence-system (*). As above, the calculations turn out to be long but trivial in each single step. Note that the attempt to verify the seemingly more natural and only slightly stronger condition $\hat{\alpha} = \alpha'\alpha$ and $\hat{\beta}_f = \beta'_f\beta_f$ fails. We emphasize that it is precisely this point which forces the seemingly unnatural definition of the stable system $\mathfrak{K}$, as presented above.)

It is clear that $\phi(id_M) = id_{\phi(M)}$ for all $M \in \hat{S}$, and that $\phi_{M,M'} : \hat{S}(M,M') \longrightarrow [\hat{C}](\phi(M),\phi(M'))$ is R-linear, for all $M,M' \in \hat{S}$.

Hence we have proved that $\phi : \hat{S} \longrightarrow [\hat{C}]$ is an R-additive functor.

2) <u>The functor</u> ϕ <u>is dense</u>. This is trivial by definition of ϕ and $[\hat{C}]$.

3) <u>The functor</u> ϕ <u>reflects isomorphisms</u>.

Let $M,M' \in \hat{S}$ and $(T,S) \in \hat{S}(M,M')$ such that $\phi(T,S)$ is an isomorphism in $[\hat{C}]$. We have to show that (T,S) is an isomorphism in $\hat{S}$. To begin with, since (T,S) is a morphism in $\hat{S}$, we have the following congruences modulo π.

$$\text{(2)} \qquad \det S \equiv \left(\prod_{i=12,14,15} \det S_{ii}\right)\cdot \det S_{99}^3 \cdot \left(\prod_{i=8,17,18} \det T_{ii}\right)\cdot \det T_{20,20}^2 \cdot \det\begin{pmatrix} S_{10,10} & S_{10,11} \\ dS_{10,11} & S_{10,10}\end{pmatrix} \cdot \det\begin{pmatrix} T_{19,19} & T_{19,26} \\ dT_{19,26} & T_{19,19}\end{pmatrix}$$

$$\text{(3)} \qquad \det T \equiv \det S_{99}^3 \cdot \det S_{12,12} \cdot \det T_{77} \cdot \det T_{17,17}^3 \cdot \det T_{20,20}^5 \cdot \left(\prod_{i=8,9,11,18} \det T_{ii}^2\right)\cdot \det\begin{pmatrix} S_{10,10} & S_{10,11} \\ dS_{10,11} & S_{10,10}\end{pmatrix} \cdot \det\begin{pmatrix} T_{16,16} & T_{16,23} \\ dT_{16,23} & T_{16,16}\end{pmatrix} \cdot \det\begin{pmatrix} T_{19,19} & T_{19,26} \\ dT_{19,26} & T_{19,19}\end{pmatrix}^2 .$$

For proving (2) and (3) one has to deduce from the congruence-system (*) sufficiently many congruences modulo π of type $S_{ij} \equiv 0$, $T_{kl} \equiv 0$, $S_{ii} \equiv S_{jj}$, $T_{kk} \equiv T_{ll}$ and $S_{ii} \equiv T_{kk}$. (Once again, this verification is a matter of patience, but it contains no mathematical difficulty.)

On the other hand, since $\phi(T,S)$ is an isomorphism in $[\hat{C}]$ we have:

(4) Each matrix in the set $\left\{ S_{ii}, T_{kk}, \begin{pmatrix} S_{10,10} & S_{10,11} \\ dS_{10,11} & S_{10,10} \end{pmatrix}, \begin{pmatrix} T_{kk} & T_{kl} \\ dT_{kl} & T_{kk} \end{pmatrix} \middle| \ i=9,12,14,15 \ ; \ k=7,8,9,11,17,18,20 \ ; \ (k,l)=(16,23) \ , \ (19,26) \right\}$ is invertible.

Therefore, combining assertions (2), (3) and (4), we see that S and T are isomorphisms in $\mathrm{mod}R_3$.

Hence, in order to show that (T,S) is an isomorphism in $\hat{S}$, it remains to prove the following property of the Krull-Schmidt subcategory K_1 of $\mathrm{mod}R_3$:

(5) If a morphism τ in K_1 is an isomorphism in $\mathrm{mod}R_3$ then τ is an isomorphism in K_1 .

PROOF of (5). Let $K, K' \in K_1$ and let $\tau \in K_1(K,K')$ be an isomorphism in $\mathrm{mod}R_3$. Then, due to the structure of the indecomposable objects of K_1 , we have that $K = K'$. Hence $\tau \in \mathrm{End}_{K_1}(K) \cap \mathrm{Aut}_{R_3}(K)$.

The block structure of τ , corresponding to the decomposition of K as an object in $\mathrm{mod}R_3$, is given by

$$\tau = \left(\begin{array}{c|c|c|c|c} \tau_{00} & 0 & \pi\tau_{02} & \pi^2\tau_{03} & \pi^2\tau_{04} \\ \hline \tau_{10} & \tau_{11} & \tau_{12} & \tau_{13} & \tau_{14} \\ \hline \tau_{20} & \pi^2 d\tau_{12} & \tau_{11} & \pi d\tau_{14} & \pi\tau_{13} \\ \hline \tau_{30} & \pi\tau_{31} & \tau_{32} & \tau_{33} & \tau_{34} \\ \hline \tau_{40} & \pi d\tau_{32} & \tau_{31} & d\tau_{34} & \tau_{33} \end{array}\right)$$

We associate with τ the morphism $\rho \in \mathrm{End}_{K_1}(K) \cap \mathrm{Aut}_{R_3}(K)$ which is given by

$$\rho = \begin{pmatrix} \tau_{00} & & & & \\ & \tau_{11} & \tau_{12} & & \\ & \pi^2 d\tau_{12} & \tau_{11} & & \\ & & & \tau_{33} & \tau_{34} \\ & & & d\tau_{34} & \tau_{33} \end{pmatrix}$$

We claim that ρ is in fact a K_1-automorphism of $K = \overset{2}{\underset{i=0}{\oplus}} K_i^{n_i}$.

Namely, let $\delta_1 = \begin{pmatrix} 0 & E \\ \pi^2 dE & 0 \end{pmatrix} \in \mathrm{End}_{K_1}(K_1^{n_1})$ and $\delta_2 = \begin{pmatrix} 0 & E \\ dE & 0 \end{pmatrix} \in \mathrm{End}_{K_1}(K_2^{n_2})$.

Then $\mathrm{End}_{K_1}(K_i^{n_i}) = \{\varepsilon_i \in \mathrm{End}_{R_3}(K_i^{n_i}) \mid \varepsilon_i\delta_i = \delta_i\varepsilon_i\}$, for all $i = 1,2$. On setting $\rho_i = \rho|K_i^{n_i}$ we obtain that $\rho_i^{-1}\delta_i = \delta_i\rho_i^{-1}$, for all $i = 1,2$. Therefore $\rho^{-1} \in \mathrm{End}_{K_1}(K)$, proving that $\rho \in \mathrm{Aut}_{K_1}(K)$. On the other hand it is easy to see that there exist K_1-automorphisms $\tau_1, \tau_2, \tau_3, \tau_4 \in \mathrm{Aut}_{K_1}(K)$ such that

$$\tau_3\tau_1\tau\tau_2\tau_4 = \begin{pmatrix} \tau_{00} & & & & \\ & \tau'_{11} & \tau_{12} & & \\ & \pi^2 d\tau_{12} & \tau'_{11} & & \\ & & & \tau'_{33} & \tau'_{34} \\ & & & d\tau'_{34} & \tau'_{33} \end{pmatrix} = \rho'$$

where $\tau'_{11} \equiv \tau_{11}$, $\tau'_{33} \equiv \tau_{33}$, $\tau'_{34} \equiv \tau_{34}$ modulo π . From $\rho \in \mathrm{Aut}_{K_1}(K)$ we deduce that $\rho' \in \mathrm{Aut}_{K_1}(K)$ and consequently that $\tau \in \mathrm{Aut}_{K_1}(K)$.

This completes the proof of (5). Hence we have proved that the functor ϕ reflects isomorphisms.

4) <u>The functor</u> ϕ <u>is full</u>.

Let $M = \sigma(X,Y,Z)$, $M' = \sigma(X',Y',Z') \in \hat{S}$ and $[\beta,\alpha] \in [\hat{C}](\phi(M),\phi(M'))$. Choose a representative $(\beta,\alpha) \in \hat{C}(\phi(M),\phi(M'))$ for $[\beta,\alpha]$. We shall construct a morphism $(T,S) \in \hat{S}(M,M')$ such that the pair of k-matrices attached to (T,S) as described in part 1 of this proof coincides with the pair of k-matrices (β_f,α) . By definition of ϕ and by Corollary 4.3 this will imply that $\phi(T,S) = [\beta,\alpha]$.

Since $(\beta,\alpha) \in \hat{C}(\phi(M),\phi(M'))$, we know that $(\alpha) = (A_{ij})_{i,j=1,\dots,8}$ and $(\beta_f) = (B_{ij})_{i,j=9,\dots,22}$ are of the form described in Lemma 4.1, (iv). In what follows we denote for any k-matrix C by $\tilde{C}$ a lifting of C to an R_i-matrix, $i = 2,3$, and we denote for any R_3-matrix D by $\bar{D}$ the reduction of D modulo π to a k-matrix.

Now we construct a pair of matrices (T,S) as indicated on pages 91-93. We have to show that the blocks S_{ij}, T_{kl} can be chosen in such a way that the equation $SM = M'T$ is satisfied. In order to do so we begin with two observations.

(6) There exists R_3-matrices $\tilde{A}_{88}$, $\tilde{B}_{99}$ and $T_{9,13}$, where $\tilde{A}_{88}$ and $\tilde{B}_{99}$ are liftings of A_{88} and B_{99} , such that $\pi\tilde{A}_{88}X_{0'} = \pi X'_{0'}\tilde{B}_{99} + \pi^2 X'_{0}T_{9,13}$.

$$
S = \begin{pmatrix}
\tilde{B}_{11,11} & d\tilde{B}_{11,16} & d\tilde{B}_{11,18} & d\tilde{B}_{11,20} & \tilde{B}_{11,19} & d\tilde{B}_{11,21} & & d\tilde{B}_{11,22} & & & & & & & \\
 & \tilde{B}_{15,15} & \tilde{B}_{16,18} & \tilde{B}_{15,19} & \tilde{B}_{15,20} & \tilde{B}_{16,21} & & \tilde{B}_{15,21} & & & & & & & \\
 & & \tilde{B}_{18,18} & & & \tilde{B}_{18,21} & & & & & & & & & \\
 & & & \tilde{B}_{19,19} & \tilde{B}_{19,20} & d\tilde{B}_{19,22} & & \tilde{B}_{19,21} & & & & & & & \\
 & & & d\tilde{B}_{19,20} & \tilde{B}_{19,19} & d\tilde{B}_{19,21} & & d\tilde{B}_{19,22} & & & & & & & \\
 & & & & & \tilde{B}_{21,21} & & & & & & & & & \\
\pi T_{22,22} & & & & & & \tilde{A}_{11} & & & \tilde{A}_{35} & d^{-1}\tilde{A}_{34} & & & & d^{-1}\tilde{A}_{18} \\
 & & & & & & & \tilde{B}_{21,21} & & & & & & & \\
S_{91} & & & & & & & & \tilde{A}_{11} & \tilde{A}_{34} & \tilde{A}_{35} & \tilde{A}_{37} & & & \tilde{A}_{38} \\
\pi T_{22,22} & \pi T_{22,1} & S_{10,3} & \pi T_{22,2} & & S_{10,6} & & & & \tilde{A}_{44} & \tilde{A}_{45} & & & & \tilde{A}_{48} \\
\pi T_{22,8} & \pi T_{29,1} & S_{11,3} & \pi T_{29,2} & \pi T_{29,3} & S_{11,6} & & \pi T_{29,4} & & d\tilde{A}_{45} & \tilde{A}_{44} & & & & \tilde{A}_{58} \\
 & & & & & & & & & & & \tilde{A}_{77} & & & \tilde{A}_{78} \\
S_{13,1} & S_{13,2} & & S_{13,4} & S_{13,5} & & & S_{13,8} & & \pi\tilde{A}_{34} & \pi\tilde{A}_{35} & & \tilde{A}_{11} & \tilde{A}_{26} & \pi\tilde{A}_{28} \\
S_{14,1} & S_{14,2} & & S_{14,4} & S_{14,5} & & & S_{14,8} & & & & & & \tilde{A}_{66} & \pi\tilde{A}_{68} \\
S_{15,1} & S_{15,2} & S_{15,3} & S_{15,4} & S_{15,5} & S_{15,6} & & S_{15,8} & & & & & & & \tilde{A}_{88}
\end{pmatrix}
$$

$$
T = \left(\begin{array}{ccccccc||cccc|cccc||}
\tilde{B}_{15,15} & \tilde{B}_{15,19} & \tilde{B}_{15,20} & \tilde{B}_{15,21} & & & -\pi T_{17,7} & & & & & & & -\pi\tilde{B}_{15,22} & -\pi\tilde{B}_{15,17} \\
& \tilde{B}_{19,19} & \tilde{B}_{19,20} & \tilde{B}_{19,21} & & & -\pi T_{19,7} & & & & & & & -\pi\tilde{B}_{19,22} & \\
& d\tilde{B}_{19,20} & \tilde{B}_{19,19} & d\tilde{B}_{19,22} & & & -\pi T_{26,7} & & & & & & & -\pi\tilde{B}_{19,21} & \\
& & & \tilde{B}_{21,21} & & & -\pi T_{27,7} & & & & & & & & \\
& & & & \tilde{A}_{11} & \tilde{A}_{37} & T_{57} & & & & & & \pi T_{5,13} & \pi T_{5,14} & \pi T_{5,15} \\
& & & & & \tilde{A}_{77} & T_{67} & & & & & & \pi T_{6,13} & \pi T_{6,14} & \pi T_{6,15} \\
-\tilde{B}_{12,15} & -\tilde{B}_{12,19} & -\tilde{B}_{12,20} & -\tilde{B}_{12,21} & & & \tilde{B}_{12,12} & & & & & & & \pi\tilde{B}_{12,22} & \pi\tilde{B}_{12,17} \\
\hline\hline
& & & & & & T_{87} & B_{11,11} & & & & & & B_{11,22} & \\
& & & & & & T_{97} & B_{9,11} & B_{99} & & & & T_{9,13} & B_{9,22} & B_{9,17} \\
& & & & & & & & & B_{21,21} & & & & & \\
& & & & & & & & & B_{17,22} & B_{17,17} & & & & \\
\hline
d\tilde{B}_{11,16} & d\tilde{B}_{11,20} & \tilde{B}_{11,19} & d\tilde{B}_{11,22} & & & & & & \pi^2 d\tilde{B}_{11,22} & & \tilde{B}_{11,11} & & & \\
T_{13,1} & T_{13,2} & T_{13,3} & -\tilde{B}_{10,21} & & & \tilde{B}_{10,12} & & \pi^2 d T_{9,13} & \pi^2 d\tilde{B}_{9,22} & \pi^2 d\tilde{B}_{9,17} & \tilde{B}_{9,11} & \tilde{B}_{99} & \pi\tilde{B}_{10,22} & \pi\tilde{B}_{10,17} \\
& & & & & & T_{14,7} & & & & & & & \tilde{B}_{21,21} & \\
& & & & & & T_{15,7} & & & & & & & \tilde{B}_{17,22} & \tilde{B}_{17,17} \\
\hline\hline
& & & & & & T_{16,7} & & & \pi\tilde{B}_{14,22} & \pi\tilde{B}_{14,17} & & & \tilde{B}_{13,22} & \tilde{B}_{13,17} \\
& & & & & & T_{17,7} & & & \pi\tilde{B}_{15,21} & & & & \tilde{B}_{15,22} & \tilde{B}_{15,17} \\
& & & & & & & & & & & & & & \\
& & & & & & T_{19,7} & & & \pi\tilde{B}_{19,21} & & & & \tilde{B}_{19,22} & \\
& & & & & & & & & & & & & & \\
& & & & & & T_{21,7} & \pi T_{21,8} & \pi T_{21,9} & & & T_{21,12} & T_{21,13} & T_{21,14} & T_{21,15} \\
T_{22,1} & T_{22,2} & & & & & T_{22,7} & \pi T_{22,8} & \pi T_{22,9} & \pi T_{22,10} & \pi T_{22,11} & T_{22,12} & T_{22,13} & T_{22,14} & T_{22,15} \\
\hline
& & & & & & T_{23,7} & & & \pi d\tilde{E}_{13,22} & \pi d\tilde{B}_{13,17} & & & \tilde{B}_{14,22} & \tilde{B}_{14,17} \\
& & & & & & T_{24,7} & & & \pi d\tilde{B}_{15,22} & \pi d\tilde{B}_{15,17} & & & \tilde{B}_{15,21} & \\
& & & & & & T_{25,7} & & & & & & & & \\
& & & & & & T_{26,7} & & & \pi d\tilde{B}_{19,22} & & & & \tilde{B}_{19,21} & \\
& & & & & & T_{27,7} & & & & & & & & \\
& & & & & & T_{28,7} & \pi d T_{21,12} & \pi d T_{21,13} & \pi d T_{21,14} & \pi d T_{21,15} & T_{21,8} & T_{21,9} & & \\
T_{29,1} & T_{29,2} & T_{29,3} & T_{29,4} & & & T_{29,7} & \pi d T_{22,12} & \pi d T_{22,13} & \pi d T_{22,14} & \pi d T_{22,15} & T_{22,8} & T_{22,9} & T_{22,10} & T_{22,11}
\end{array}\right.
$$

$\pi^2 T_{5,16}$	$\pi^2 T_{5,17}$	$\pi^2 T_{6,18}$	$\pi^2 T_{5,19}$	$\pi^2 T_{5,20}$			$\pi^2 T_{5,23}$	$\pi^2 T_{5,24}$	$\pi^2 T_{5,25}$	$\pi^2 T_{5,26}$	$\pi^2 T_{5,27}$		
$\pi^2 T_{6,16}$	$\pi^2 T_{6,17}$	$\pi^2 T_{6,18}$	$\pi^2 T_{6,19}$	$\pi^2 T_{6,20}$			$\pi^2 T_{6,23}$	$\pi^2 T_{6,24}$	$\pi^2 T_{6,25}$	$\pi^2 T_{6,26}$	$\pi^2 T_{6,27}$		
			$B_{11,19}$	$dB_{11,22}$				$B_{11,16}$	$B_{11,18}$	$B_{11,20}$	$B_{11,21}$		
$B_{9,13}$	$B_{9,15}$	$B_{10,18}$	$B_{9,19}$				$B_{9,14}$	$B_{9,16}$	$B_{9,18}$	$B_{9,20}$	$B_{9,21}$		
	$\pi d\tilde{B}_{11,16}$	$\pi d\tilde{B}_{11,18}$	$\pi d\tilde{B}_{11,20}$	$\pi d\tilde{B}_{11,21}$						$\pi\tilde{B}_{11,19}$	$\pi d\tilde{B}_{11,22}$		
$\pi d\tilde{B}_{9,14}$	$\pi d\tilde{B}_{9,16}$	$\pi d\tilde{B}_{9,18}$	$\pi d\tilde{B}_{9,20}$	$\pi d\tilde{B}_{9,21}$			$\pi\tilde{B}_{9,13}$	$\pi\tilde{B}_{9,15}$	$\pi\tilde{B}_{10,18}$	$\pi\tilde{B}_{9,19}$			
$\tilde{B}_{13,13}$	$\tilde{B}_{13,15}$	$\tilde{B}_{14,18}$	$\tilde{B}_{13,19}$	$\tilde{B}_{14,21}$			$\tilde{B}_{13,14}$	$\tilde{B}_{13,16}$	$\tilde{B}_{13,18}$	$\tilde{B}_{13,20}$	$\tilde{B}_{13,21}$		
	$\tilde{B}_{15,15}$	$\tilde{B}_{16,18}$	$\tilde{B}_{15,19}$	$\tilde{B}_{16,21}$						$\tilde{B}_{15,20}$	$\tilde{B}_{15,21}$		
		$\tilde{B}_{18,18}$		$\tilde{B}_{18,21}$									
			$\tilde{B}_{19,19}$	$d\tilde{B}_{19,22}$						$\tilde{B}_{19,20}$	$\tilde{B}_{19,21}$		
				$\tilde{B}_{21,21}$									
$T_{21,16}$	$T_{21,17}$		$T_{21,19}$		$\tilde{A}_{11}$	$\tilde{A}_{35}$	$T_{21,23}$	$T_{21,24}$	$T_{21,25}$	$T_{21,26}$	$T_{21,27}$		$d^{-1}\tilde{A}_{34}$
$T_{22,16}$	$T_{22,17}$	$T_{22,18}$				$\tilde{A}_{44}$	$T_{22,23}$	$T_{22,24}$	$T_{22,25}$	$T_{22,26}$	$T_{22,27}$		$\tilde{A}_{45}$
$d\tilde{B}_{13,14}$	$d\tilde{B}_{13,16}$	$d\tilde{B}_{13,18}$	$d\tilde{B}_{13,20}$	$d\tilde{B}_{13,21}$			$\tilde{B}_{13,13}$	$\tilde{B}_{13,15}$	$\tilde{B}_{14,18}$	$\tilde{B}_{13,19}$	$\tilde{B}_{14,21}$		
			$d\tilde{B}_{15,20}$	$d\tilde{B}_{15,21}$				$\tilde{B}_{15,15}$	$\tilde{B}_{16,18}$	$\tilde{B}_{15,19}$	$\tilde{B}_{16,21}$		
									$\tilde{B}_{18,18}$		$\tilde{B}_{18,21}$		
			$d\tilde{B}_{19,20}$	$d\tilde{B}_{19,21}$						$\tilde{B}_{19,19}$	$d\tilde{B}_{19,22}$		
											$\tilde{B}_{21,21}$		
$dT_{21,23}$	$dT_{21,24}$	$dT_{21,25}$	$dT_{21,26}$	$dT_{21,27}$		$\tilde{A}_{34}$	$T_{21,16}$	$T_{21,17}$		$T_{21,19}$		$\tilde{A}_{11}$	$\tilde{A}_{35}$
$dT_{22,23}$	$dT_{22,24}$	$dT_{22,25}$	$dT_{22,26}$	$dT_{22,27}$		$d\tilde{A}_{45}$	$T_{22,16}$	$T_{22,17}$	$T_{22,18}$				$\tilde{A}_{44}$

(7) There exist R_3-matrices $\tilde{B}_{10,12}$, $\tilde{B}_{12,12}$ and T_{i7} , $i \in \{8,9,14,15,16,17,19,23,24,25,26,27\}$, where $\tilde{B}_{10,12}$ and $\tilde{B}_{12,12}$ are liftings of $B_{10,12}$ and $B_{12,12}$, such that $\pi\tilde{A}_{88}X_2 = \pi(X'_{0'}\tilde{B}_{10,12}+X'_2\tilde{B}_{10,12}) + \pi^2(X'_0T_{97}+X'_1T_{87}+X'_3T_{16,7}+X'_{3'}T_{23}+X'_4T_{17,7}+X'_{4'}T_{24,7}+$ $+X'_5T_{15,7}+X'_6T_{25,7}+X'_7T_{19,7}+X'_{7'}T_{26,7}+X'_8T_{27,7}+X'_{8'}T_{14,7})$.

PROOF of (6). Due to the second reduction (proof of Proposition 2.1) we can assume that $(\pi^2X_0|\pi X_{0'}) = \left(\begin{array}{c|c|c|c} 0 & 0 & 0 & 0 \\ \hline \pi^2K_1 & 0 & \pi K_r & 0 \end{array}\right)$ and $(\pi^2X'_0|\pi X'_{0'}) = \left(\begin{array}{c|c|c|c} 0 & 0 & 0 & 0 \\ \hline \pi^2K'_1 & 0 & \pi K'_r & 0 \end{array}\right)$, such that $(\bar{K}_1|\bar{K}_r)$, $(\bar{K}'_1|\bar{K}'_r)$ correspond to Kronecker modules which contain no semisimple direct summand. In other words, the k-linear maps corresponding to $(\bar{K}_1|\bar{K}_r)$ and $(\bar{K}'_1|\bar{K}'_r)$ are epimorphisms and the k-linear maps corresponding to $\begin{pmatrix} \bar{K}_1 \\ \bar{K}_r \end{pmatrix}$ and $\begin{pmatrix} \bar{K}'_1 \\ \bar{K}'_r \end{pmatrix}$ are monomorphisms. Now put $A_{88} = \begin{pmatrix} \alpha_1 & \alpha_2 \\ \alpha_3 & \alpha_4 \end{pmatrix}$ and $B_{99} = \begin{pmatrix} \beta_1 & \beta_2 \\ \beta_3 & \beta_4 \end{pmatrix}$, according to the block partitions of $(\bar{X}_0|\bar{X}_{0'})$ and $(\bar{X}'_0|\bar{X}'_{0'})$. Then the equations $A_{88}\bar{X}_0 = \bar{X}'_0B_{99}$, $A_{88}\bar{X}_{0'} = \bar{X}'_{0'}B_{99}$ (which are valid since $(\beta,\alpha) \in \hat{C}(\phi(M),\phi(M'))$ imply that $\alpha_2 = 0$, $\beta_2 = 0$ and $\alpha_4\bar{K}_r = \bar{K}'_r\beta_1$. Choose R_3-matrices $\tilde{A}_{88} = \begin{pmatrix} \tilde{\alpha}_1 & 0 \\ \tilde{\alpha}_3 & \tilde{\alpha}_4 \end{pmatrix}$ and $\hat{B}_{99} = \begin{pmatrix} \tilde{\beta}_1 & 0 \\ \tilde{\beta}_3 & \tilde{\beta}_4 \end{pmatrix}$ which are liftings of A_{88} and B_{99} . Then $\pi\tilde{A}_{88}X_{0'} = \pi X'_{0'}\hat{B}_{99} + \begin{pmatrix} 0 & 0 \\ \pi^2C & 0 \end{pmatrix}$, for some R_3-matrix C . Since the k-linear map corresponding to $(\bar{K}'_1\bar{K}'_r)$ is an epimorphism, there exists an R_3-matrix $\begin{pmatrix} D_1 \\ D_r \end{pmatrix}$ such that $\pi^2(K_1\ K_r)\begin{pmatrix} D_1 \\ D_r \end{pmatrix} = \pi^2C$. Therefore the R_3-matrices $\tilde{A}_{88}$, $\tilde{B}_{99} = \hat{B}_{99} +$

$+ \begin{pmatrix} \pi D_r & 0 \\ 0 & 0 \end{pmatrix}$, $T_{9,13} = \begin{pmatrix} D_1 & 0 \\ 0 & 0 \end{pmatrix}$ satisfy the equation $\pi\tilde{A}_{88}X_{0'} = \pi X'_{0'}\tilde{B}_{99} + \pi^2 X'_0 T_{9,13}$. This proves assertion (6).

PROOF of (7). Let $\hat{B}_{10,12}$ and $\hat{B}_{12,12}$ be R_3-matrices which are liftings of $B_{10,12}$ and $B_{12,12}$. Since by assumption $(\beta,\alpha) \in C(\phi(M),\phi(M'))$, we have that $A_{88}\bar{X}_2 = \bar{X}'_{0'}B_{10,12}+\bar{X}'_2 B_{12,12}$, hence $\pi\tilde{A}_{88}X_2 = \pi(X'_{0'}\hat{B}_{10,12}+X'_2\hat{B}_{12,12}) + \pi^2 C$, for some R_3-matrix C . Also by assumption $M' = \sigma(X',Y',Z') \in \hat{S}$ where, by definition of $\hat{S}$, $(X',Y',Z') \in \mathfrak{M}$. Now it is straightforward to deduce from the definition of $\mathfrak{M}$ that the k-linear map corresponding to $\bar{X}' = (\bar{X}'_i)_{i\in I_X}$ is an epimorphism. Hence there exist R_3-matrices T_{i7} , $i \in \{7,8,9,13,14,15,16,17,19,23,24,25,26,27\}$, such that $\pi^2 C = \pi^2(X'_0 T_{97} + X'_{0'}T_{13,7} + X'_1 T_{87} + X'_2 T_{77} + X'_3 T_{16,7} + X'_{3'}T_{23,7} + X'_4 T_{17,7} + X'_{4'}T_{24,7} + X'_5 T_{15,7} + X'_6 T_{25,7} + X'_7 T_{19,7} + X'_{7'}T_{26,7} + X'_8 T_{27,7} + X'_{8'}T_{14,7})$. On setting $\tilde{B}_{10,12} = \hat{B}_{10,12} + \pi T_{13,7}$ and $\tilde{B}_{12,12} = \hat{B}_{12,12} + \pi T_{77}$, this proves assertion (7).

Now we are ready to complete the construction of the morphism $(T,S) \in \hat{S}(M,M')$ by specifying the blocks which constitute the matrices S and T given above. We begin with several choices. First of all, choose $\tilde{A}_{88}$, $\tilde{B}_{99}$, $T_{9,13}$ according to assertion (6). Next, choose $\tilde{B}_{10,12}$, $\tilde{B}_{12,12}$, T_{i7} $(i \in \{8,9,14,15,16,17,19,23,24,25,26,27\})$ according to assertion (7). Moreover, choose all remaining blocks $\tilde{A}_{ij}$, $\tilde{B}_{kl}$ to be arbitrary liftings of A_{ij} and B_{kl} .

After these choices it turns out that the equation $SM = M'T$, viewed as a system of 435 linear congruences in the remaining blocks S_{ij} and T_{kl}, has a uniquely determined solution which is given as follows.

$$
\begin{aligned}
S_{91} &= \pi(d^{-1}\tilde{A}_{18}X_1 + \pi Y'_{4'}\tilde{A}_{37}) , \\
S_{10,3} &= T_{22,18} + \pi(Y'_{3'}\tilde{B}_{10,18} + Z'_{3'3}\tilde{B}_{14,18} + dZ'_{3'3'}\tilde{B}_{13,18}) , \\
S_{10,6} &= \pi(Z'_{3'3}\tilde{B}_{14,21} + dZ'_{3'3'}\tilde{B}_{13,21}) , \\
S_{11,3} &= \pi dT_{22,25} + \pi^2(Y'_3\tilde{B}_{10,18} + Z'_{33}\tilde{B}_{14,18} + dZ'_{33'}\tilde{B}_{13,18}) , \\
S_{11,6} &= \pi dT_{22,27} + Z'_{33}\tilde{B}_{14,21} + dZ'_{33'}\tilde{B}_{13,21} , \\
S_{13,1} &= \pi^2(T_{21,8} + Y'_4\tilde{B}_{9,11}) , \\
S_{13,2} &= \pi^2 Y'_4 T_{13,1} , \\
S_{13,4} &= \pi^2 Y'_4 T_{13,2} , \\
S_{13,5} &= \pi^2 Y'_4 T_{13,3} , \\
S_{13,8} &= -\pi^2 Y'_4\tilde{B}_{10,21} , \\
S_{14,1} &= \pi^2 Y'_2\tilde{B}_{9,11} , \\
S_{14,2} &= \pi^2(Y'_2 T_{13,1} - Z'_2\tilde{B}_{12,15}) , \\
S_{14,4} &= \pi^2(Y'_2 T_{13,2} - Z'_2\tilde{B}_{12,19}) , \\
S_{14,5} &= \pi^2(Y'_2 T_{13,3} - Z'_2\tilde{B}_{12,20}) , \\
S_{14,8} &= -\pi^2(Y'_2\tilde{B}_{10,21} + Z'_2\tilde{B}_{12,21}) , \\
S_{15,1} &= \pi X'_{0'}\tilde{B}_{9,11} , \\
S_{15,2} &= \pi(X'_{0'}T_{13,1} - X'_2\tilde{B}_{12,15}) , \\
S_{15,3} &= \pi(X'_0\tilde{B}_{10,18} + X'_{0'}d\tilde{B}_{9,18} + X'_3\tilde{B}_{14,18} + X'_{3'}d\tilde{B}_{13,18} + X'_4\tilde{B}_{16,18}) , \\
S_{15,4} &= \pi(X'_{0'}T_{13,2} - X'_2\tilde{B}_{12,19}) , \\
S_{15,5} &= \pi(X'_{0'}T_{13,3} - X'_2\tilde{B}_{12,20}) , \\
S_{15,6} &= \pi(X'_{0'}d\tilde{B}_{9,21} + X'_1 d\tilde{B}_{11,22} + X'_3\tilde{B}_{14,21} + X'_{3'}d\tilde{B}_{13,21} + X'_4\tilde{B}_{16,21} + \\
&\quad + X'_{4'}d\tilde{B}_{15,21} + X'_7 d\tilde{B}_{19,22} + X'_{7'}d\tilde{B}_{19,21}) , \\
S_{15,8} &= -\pi(X'_{0'}\tilde{B}_{10,21} + X'_2\tilde{B}_{12,21}) ,
\end{aligned}
$$

$$T_{57} = \pi(\tilde{A}_{38}X_2 - T_{28,7} - \pi Y'_{4'}T_{97}) ,$$

$$T_{5,13} = \tilde{A}_{38}X_{0'} - T_{21,9} - \pi Y'_{4'}T_{9,13} ,$$

$$T_{5,14} = \pi(\tilde{A}_{38}X_{8'} - Y'_{4'}\tilde{B}_{9,22}) ,$$

$$T_{5,15} = \pi(\tilde{A}_{38}X_5 - Y'_{4'}\tilde{B}_{9,17}) ,$$

$$T_{5,16} = \tilde{A}_{34}Z_{3'3} + \tilde{A}_{35}Z_{33} + \tilde{A}_{38}X_3 - Y'_4\tilde{B}_{9,13} - (d\tilde{A}_{35}Z_{3'3'} + \tilde{A}_{34}Z_{33'} + \tilde{A}_{18}X_{3'}),$$

$$T_{5,17} = \tilde{A}_{38}X_4 - Y'_{4'}\tilde{B}_{9,15} - \tilde{A}_{18}X_{4'} ,$$

$$T_{5,18} = -(Y'_{4'}\tilde{B}_{10,18} + \tilde{A}_{18}X_6) ,$$

$$T_{5,19} = \tilde{A}_{38}X_7 - (Y'_{4'}\tilde{B}_{9,19} + \tilde{A}_{18}X_{7'}) ,$$

$$T_{5,20} = -\tilde{A}_{18}X_8 ,$$

$$T_{5,23} = \tilde{A}_{34}Z_{3'3'} + \tilde{A}_{35}Z_{33'} + \tilde{A}_{38}X_{3'} - (Y'_{4'}\tilde{B}_{9,14} + \tilde{A}_{35}Z_{3'3} + d^{-1}\tilde{A}_{34}Z_{33} + d^{-1}\tilde{A}_{18}X_3),$$

$$T_{5,24} = \tilde{A}_{38}X_{4'} - (Y'_{4'}\tilde{B}_{9,16} + d^{-1}\tilde{A}_{18}X_4) ,$$

$$T_{5,25} = \tilde{A}_{38}X_6 - Y'_{4'}\tilde{B}_{9,18} ,$$

$$T_{5,26} = \tilde{A}_{38}X_{7'} - (Y'_{4'}\tilde{B}_{9,20} + d^{-1}\tilde{A}_{18}X_7) ,$$

$$T_{5,27} = \tilde{A}_{38}X_8 - Y'_{4'}\tilde{B}_{9,21} ,$$

$$T_{67} = \pi(\tilde{A}_{78}X_2 - Z'_1T_{87} - Y'_1T_{97}) ,$$

$$T_{6,13} = A_{78}X_{0'} - \pi Y'_1T_{9,13} ,$$

$$T_{6,14} = \pi(\tilde{A}_{78}X_{8'} - Z'_1\tilde{B}_{11,22} - Y'_1\tilde{B}_{9,22}) ,$$

$$T_{6,15} = \pi(\tilde{A}_{78}X_5 - Y'_1\tilde{B}_{9,17}) ,$$

$$T_{6,16} = \tilde{A}_{78}X_3 - Y'_1\tilde{B}_{9,13} ,$$

$$T_{6,17} = \tilde{A}_{78}X_4 - Y'_1\tilde{B}_{9,15} ,$$

$$T_{6,18} = -Y'_1\tilde{B}_{10,18} ,$$

$$T_{6,19} = \tilde{A}_{78}X_7 - Z'_1\tilde{B}_{11,19} - Y'_1\tilde{B}_{9,19} ,$$

$$T_{6,20} = -Z'_1d\tilde{B}_{11,22} ,$$

$$T_{6,23} = \tilde{A}_{78}X_{3'} - Y'_1\tilde{B}_{9,14} ,$$

$$T_{6,24} = \tilde{A}_{78}X_{4'} - Z'_1\tilde{B}_{11,16} - Y'_1\tilde{B}_{9,16} ,$$

$$T_{6,25} = \tilde{A}_{78}X_6 - Z'_1\tilde{B}_{11,18} - Y'_1\tilde{B}_{9,18} ,$$

$$
\begin{aligned}
T_{6,26} &= \tilde{A}_{78}X_{7'} - Z'_1\tilde{B}_{11,20} - Y'_1\tilde{B}_{9,20} , \\
T_{6,27} &= \tilde{A}_{78}X_8 - Z'_1\tilde{B}_{11,21} - Y'_1\tilde{B}_{9,21} , \\
T_{13,1} &= d\tilde{B}_{9,16} - \tilde{B}_{10,15} , \\
T_{13,2} &= d\tilde{B}_{9,20} - \tilde{B}_{10,19} , \\
T_{13,3} &= \tilde{B}_{9,19} - \tilde{B}_{10,20} , \\
T_{21,7} &= d^{-1}\tilde{A}_{18}X_2 , \\
T_{21,8} &= d^{-1}\tilde{A}_{18}X_1 , \\
T_{21,9} &= \tilde{A}_{35}Y_{3'} + d^{-1}\tilde{A}_{34}Y_3 + d^{-1}\tilde{A}_{18}X_0 , \\
T_{21,12} &= d^{-1}(\tilde{A}_{37}Z_1 + \tilde{A}_{38}X_1 - Y'_{4'}\tilde{B}_{9,11}) , \\
T_{21,13} &= d^{-1}\tilde{A}_{18}X_{0'} , \\
T_{21,14} &= \pi d^{-1}\tilde{A}_{18}X_{8'} , \\
T_{21,15} &= \pi d^{-1}\tilde{A}_{18}X_5 , \\
T_{21,16} &= \pi(\tilde{A}_{35}Z_{3'3} + d^{-1}\tilde{A}_{34}Z_{33} + d^{-1}\tilde{A}_{18}X_3) , \\
T_{21,17} &= \pi d^{-1}\tilde{A}_{18}X_4 , \\
T_{21,19} &= \pi d^{-1}\tilde{A}_{18}X_7 , \\
T_{21,23} &= \pi(\tilde{A}_{35}Z_{3'3'} + d^{-1}\tilde{A}_{34}Z_{33'} + d^{-1}\tilde{A}_{18}X_{3'}) , \\
T_{21,24} &= \pi d^{-1}\tilde{A}_{18}X_{4'} , \\
T_{21,25} &= \pi d^{-1}\tilde{A}_{18}X_6 , \\
T_{21,26} &= \pi d^{-1}\tilde{A}_{18}X_{7'} , \\
T_{21,27} &= \pi d^{-1}\tilde{A}_{18}X_8 , \\
T_{22,1} &= -\tilde{A}_{48}X_4 + \tilde{A}_{58}X_{4'} - Y'_3\tilde{B}_{9,16} + Y'_{3'}\tilde{B}_{9,15} - Z'_{33}\tilde{B}_{13,16} - Z'_{33'}\tilde{B}_{13,15} + \\
&\quad + Z'_{3'3}\tilde{B}_{13,15} + dZ'_{3'3'}\tilde{B}_{13,16} , \\
T_{22,2} &= -\tilde{A}_{48}X_7 + Y'_{3'}\tilde{B}_{9,19} + Z'_{3'3}\tilde{B}_{13,19} + dZ'_{3'3'}\tilde{B}_{13,20} , \\
T_{22,7} &= \tilde{A}_{48}X_2 - \pi(Y'_{3'}T_{97} + Z'_{3'3}T_{16,7} + Z'_{3'3'}T_{23,7}) , \\
T_{22,8} &= \tilde{A}_{48}X_1 - Y'_{3'}\tilde{B}_{9,11} , \\
T_{22,9} &= \tilde{A}_{58}X_{0'} - \pi Y'_3T_{9,13} ,
\end{aligned}
$$

$T_{22,10} = \pi(\tilde{A}_{58}X_{8'} - Y'_3\tilde{B}_{9,22} - Z'_{33}\tilde{B}_{13,22} - Z'_{33'}\tilde{B}_{14,22})$,

$T_{22,11} = \pi(\tilde{A}_{58}X_5 - Y'_3\tilde{B}_{9,22} - Z'_{33}\tilde{B}_{13,17} - Z'_{33'}\tilde{B}_{14,17})$,

$T_{22,12} = d^{-1}(\tilde{A}_{58}X_1 - Y'_3\tilde{B}_{9,11})$,

$T_{22,13} = \tilde{A}_{48}X_{0'} - \pi Y'_{3'}T_{9,13}$,

$T_{22,14} = \pi(\tilde{A}_{48}X_{8'} - Y'_{3'}\tilde{B}_{9,22} - Z'_{3'3}\tilde{B}_{13,22} - Z'_{3'3'}\tilde{B}_{14,22})$,

$T_{22,15} = \pi(\tilde{A}_{48}X_5 - Y'_{3'}\tilde{B}_{9,17} - Z'_{3'3}\tilde{B}_{13,17} - Z'_{3'3'}\tilde{B}_{14,17})$,

$T_{22,16} = \pi(\tilde{A}_{44}Z_{3'3}+\tilde{A}_{45}Z_{33}+\tilde{A}_{48}X_3-Y'_{3'}\tilde{B}_{9,13}-Z'_{3'3}\tilde{B}_{13,13}-Z'_{3'3'}d\tilde{B}_{13,14})$,

$T_{22,17} = \pi(\tilde{A}_{58}X_{4'} - Y'_3\tilde{B}_{9,16} - Z'_{33}\tilde{B}_{13,16} - Z'_{33'}\tilde{B}_{13,15})$,

$T_{22,18} = \pi(\tilde{A}_{58}X_6 - Y'_3\tilde{B}_{9,18} -Z'_{33}\tilde{B}_{13,18} - Z'_{33'}\tilde{B}_{14,18})$,

$T_{22,23} = \pi(\tilde{A}_{44}Z_{3'3'}+\tilde{A}_{45}Z_{33'}+\tilde{A}_{48}X_{3'}-Y'_{3'}\tilde{B}_{9,14}+Z'_{3'3}\tilde{B}_{13,14}+Z'_{3'3'}\tilde{B}_{13,13})$,

$T_{22,24} = \pi(\tilde{A}_{48}X_{4'} - Y'_{3'}\tilde{B}_{9,16} - Z'_{3'3}\tilde{B}_{13,16} - Z'_{3'3'}\tilde{B}_{13,15})$,

$T_{22,25} = \pi(\tilde{A}_{48}X_6 - Y'_{3'}\tilde{B}_{9,18} - Z'_{3'3}\tilde{B}_{13,18} - Z'_{3'3'}\tilde{B}_{14,18})$,

$T_{22,26} = \pi(\tilde{A}_{48}X_{7'} - Y'_{3'}\tilde{B}_{15,20} - Z'_{3'3}\tilde{B}_{13,20} - Z'_{3'3'}\tilde{B}_{13,19})$,

$T_{22,27} = \pi(\tilde{A}_{48}X_8 - Y'_{3'}\tilde{B}_{9,21} - Z'_{3'3}\tilde{B}_{13,21} - Z'_{3'3'}\tilde{B}_{14,21})$,

$T_{28,7} = \tilde{A}_{28}X_2 + \tilde{A}_{26}Z_2 - Y'_4\tilde{B}_{10,12}$,

$T_{29,1} = d(\tilde{A}_{48}X_{4'} -Y'_{3'}\tilde{B}_{9,16} - Z'_{3'3}\tilde{B}_{13,16} - Z'_{3'3'}\tilde{B}_{13,15}) + Y'_3\tilde{B}_{9,15} +$
$+ Z'_{33}\tilde{B}_{13,15} + Z'_{33'}d\tilde{B}_{13,16} - \tilde{A}_{58}X_4$,

$T_{29,2} = d(\tilde{A}_{48}X_{7'} - Y'_{3'}\tilde{B}_{15,20} - Z'_{3'3}\tilde{B}_{13,20} - Z'_{3'3'}\tilde{B}_{13,19}) + Y'_3\tilde{B}_{9,19} +$
$+ Z'_{33}\tilde{B}_{13,19} + Z'_{33'}d\tilde{B}_{13,20} - \tilde{A}_{58}X_7$,

$T_{29,3} = Y'_3\tilde{B}_{9,20} + Z'_{33}\tilde{B}_{13,20} + Z'_{33'}\tilde{B}_{13,19} - \tilde{A}_{58}X_{7'}$,

$T_{29,4} = Y'_3\tilde{B}_{9,21} + Z'_{33}\tilde{B}_{13,21} + Z'_{33'}\tilde{B}_{14,21} - \tilde{A}_{58}X_8$,

$T_{29,7} = \tilde{A}_{58}X_2 - \pi(Y'_3T_{97} + Z'_{33}T_{16,7} + Z'_{33'}T_{23,7})$.

This finishes our construction of the matrices S and T . Now it is a matter of patience to verify that $(T,S) \in \hat{S}(M,M')$ and $\phi(T,S) = [\beta,\alpha]$. Hence we have shown that the functor ϕ is full.

This completes the proof. q.e.d.

COROLLARY 4.5. If all components of $A([\hat{C}])$ are stable tubes then the Auslander-Reiten quivers $A(\hat{S})$ and $A([\hat{C}])$ are isomorphic.

PROOF. Let R be the trivial indecomposable Λ-lattice, and let M be an arbitrary indecomposable nonprojective Λ-lattice. It is easily seen that there is an exact sequence $0 \to R \to \Lambda \to \Lambda \to R \to 0$ in ${}_{\Lambda}L$. Tensoring with M shows that $\Omega^2 M \cong M$, for each indecomposable nonprojective Λ-lattice M, where Ω denotes Heller's operator. Since Λ is a Gorenstein order of dimension 1 we have $\tau = \Omega$, by [Au 76]. Hence every point in $A_s(\Lambda)$ is τ-periodic of period 1 or 2. Since in addition Λ is of infinite representation type, every component of $A_s(\Lambda)$ is a stable tube of rank 1 or 2 [Wi 80]. In view of the first, second and third reduction (Corollaries 1.3, 2.2 and 3.6) we conclude that all components of $A(\hat{S})$ are stable tubes of rank 1 or 2.

On the other hand, the representation equivalence $\phi : \hat{S} \to [\hat{C}]$ induces a bijection between the sets of points of $A(\hat{S})$ and of $A([\hat{C}])$, and it induces bimodule epimorphisms $\varepsilon_{M,M'} : \mathrm{irr}_{\hat{S}}(M,M') \twoheadrightarrow \mathrm{irr}_{[\hat{C}]}(\phi(M),\phi(M'))$ for all $M,M' \in \mathrm{ind}\hat{S}$. Because all arrows in $A(\hat{S})$ have valuation (1,1), all epimorphisms $\varepsilon_{M,M'}$ are either isomorphisms or zero morphisms.

Now suppose that all components of $A([\hat{C}])$ are stable tubes. Since we also know that all components of $A(\hat{S})$ are stable tubes of rank 1 or 2, straightforward combinatorial arguments show that $\varepsilon_{M,M'}$ is an isomorphism for all $M,M' \in \mathrm{ind}\hat{S}$. This proves the isomorphism of quivers $A(\hat{S}) \cong A([\hat{C}])$. q.e.d.

LEMMA 4.6. The Auslander-Reiten quivers $A([\hat{C}])$ and $A(\hat{C})$ are isomorphic.

PROOF. Let $\Psi : \hat{C} \longrightarrow [\hat{C}]$ be the canonical functor. (It is given by $\Psi(X) = X$ on objects and by $\Psi(\beta,\alpha) = [\beta,\alpha] = (\beta,\alpha) + \mathfrak{K}(X,X')$ on morphisms.) By definition, Ψ is dense and full. We claim that Ψ has the following two additional properties:

(1) Ψ reflects isomorphisms.

(2) $\mathfrak{K}(X,X') \subset \mathrm{rad}^2_{\hat{C}}(X,X')$, for all $X,X' \in \mathrm{ind}\hat{\mathcal{C}}$.

Of course, if this claim is true then Ψ induces an isomorphism of Auslander-Reiten quivers $A(\hat{C}) \cong A([\hat{C}])$. Hence it remains to prove (1) and (2).

Preparatory to this we have to investigate the morphisms $(\rho,0)$ in $\mathfrak{K}(X,X')$ for arbitrary objects $X = (K,F,\varphi)$, $X' = (K',F',\varphi') \in \hat{C}$. We may write $K = \bigoplus_{i=-3}^{8} K_i^{n_i}$, $K' = \bigoplus_{i=-3}^{8} K_i^{n'_i}$, $\varphi = \begin{pmatrix}\varphi_u & \varphi_e\\ 0 & \varphi_f\end{pmatrix}$, $\varphi' = \begin{pmatrix}\varphi'_u & \varphi'_e\\ 0 & \varphi'_f\end{pmatrix}$, $\rho = \begin{pmatrix}\rho_u & \rho_e\\ 0 & \rho_f\end{pmatrix}$. From Lemma 4.1 (ii) we conclude that $\rho_u = 0$. In other words, $\bigoplus_{i=-3}^{-1} K_i^{n_i} \subset \ker\rho$. Moreover, $\varphi'\rho = 0$ implies that $\varphi'_u\rho_e = -\varphi'_e\rho_f$. Taking into account that (φ'_u) is a matrix in given normal form and that (φ'_e), (ρ_f) and (ρ_e) are matrices with prescribed block structure, an evaluation of the last equation shows that $\bigoplus_{i=0}^{3} K_i^{n_i} \subset \ker\rho$. Altogether we see that $\bigoplus_{i=-3}^{3} K_i^{n_i} \subset \ker\rho$, and $\mathrm{im}\rho \subset \bigoplus_{i=-3}^{0} K_i^{n'_i}$. (The latter inclusion follows immediately from the definition of $\mathfrak{K}(X,X')$.) In particular, if $(\rho,0) \in \mathfrak{K}(X,X)$ then $\rho^2=0$.

PROOF of (1). Let $X = (K,F,\varphi)$, $X' = (K',F',\varphi') \in \hat{C}$, and let $(\beta,\alpha) \in \hat{C}(X,X')$ such that $\Psi(\beta,\alpha) = [\beta,\alpha]$ is an isomorphism in $[\hat{C}]$.

Set $[\delta,\gamma] = [\beta,\alpha]^{-1}$. Then $(\delta,\gamma)(\beta,\alpha) = (1_K+\rho, 1_F)$ and $(\beta,\alpha)(\delta,\gamma) = (1_{K'}+\rho', 1_{F'})$, where $(\rho,0) \in \mathfrak{K}(X,X)$ and $(\rho',0) \in \mathfrak{K}(X',X')$. Because $\rho^2 = 0$ and $(\rho')^2 = 0$, we obtain that $((1_K-\rho)\delta,\gamma)(\beta,\alpha) = (1_K, 1_F)$ and $(\beta,\alpha)(\delta(1_{K'}-\rho'),\gamma) = (1_{K'}, 1_{F'})$. Hence (β,α) is an isomorphism in $\hat{C}$.

PROOF of (2). Let $X = (K,F,\varphi)$, $X' = (K',F',\varphi') \in \mathrm{ind}\hat{C}$, and let $(\rho,0) \in \mathfrak{K}(X,X')$. We have to show that $(\rho,0) \in \mathrm{rad}^2_{\hat{C}}(X,X')$.

Set $K_* = \bigoplus_{i=4}^{8} K_i^{n_i}$ and $Z = (K_*,0,0)$. Consider the canonical projection $\rho_1 : K \twoheadrightarrow K_*$ and the restriction $\rho_2 = \rho|_{K_*} : K_* \longrightarrow K'$. Then the inclusion $\bigoplus_{i=-3}^{3} K_i^{n_i} \subset \ker\rho$ gives rise to a factorization $(\rho,0) = (\rho_2,0)(\rho_1,0)$ of $(\rho,0)$ over Z , and Z is in $\hat{C}$. The factor $(\rho_2,0)$ is in $\mathrm{rad}_{\hat{C}}(Z,X')$, because $\mathrm{im}\rho_2 = \mathrm{im}\rho \subset \bigoplus_{i=-3}^{0} K_i^{n'_i}$. Hence, if $(\rho_1,0) \in \mathrm{rad}_{\hat{C}}(X,Z)$ then assertion (2) is true.

Therefore we may assume that $(\rho_1,0) \notin \mathrm{rad}_{\hat{C}}(X,Z)$. Consequently $X = (K_i,0,0)$, with $i \geq 4$. Set $K'_* = \bigoplus_{i=-3}^{0} K_i^{n'_i}$, $\rho'_1 = \rho : K \longrightarrow K'_*$, $\rho'_2 : K'_* \longrightarrow K'$ (the inclusion mapping) , and $Z' = (K'_*,F',\varphi'\rho'_2)$. Moreover, let $Z' = \check{Z}' \oplus (0,k,0)^{\ell}$ such that $\check{Z}' = (K'_*,U',\varphi'\rho'_2)$ does not contain a direct summand of type $(0,k,0)$, and let $\iota' : U' \longrightarrow F'$ be the inclusion mapping. Then the inclusion $\mathrm{im}\rho \subset K'_*$ gives rise to a factorization $(\rho,0) = (\rho'_2,\iota')(\rho'_1,0)$ of $(\rho,0)$ over $\check{Z}'$, and $\check{Z}'$ is in $\hat{C}$. Evidently the factor $(\rho'_1,0)$ is in $\mathrm{rad}_{\hat{C}}(X,\check{Z}')$. Hence, if $(\rho'_2,\iota') \in \mathrm{rad}_{\hat{C}}(\check{Z}',X')$ then assertion (2) is true.

Therefore we may assume that $(\rho'_2,\iota') \notin \mathrm{rad}_{\hat{C}}(\check{Z}',X')$. Consequently $X' = (K',F',\varphi')$, with $K' = \bigoplus_{i=-3}^{0} K_i^{n'_i}$. By definition of the category K , there exists an object $K''_* = \bigoplus_{i=1}^{3} K_i^{m_i}$ together with morphisms

$\rho_1'' : K \longrightarrow K_*''$ and $\rho_2'' : K_*'' \longrightarrow K'$ such that $\rho = \rho_2''\rho_1''$. Set $Z'' = (K_*'', F', \varphi'\rho_2'')$ and let $Z'' = \check{Z}'' \oplus (0,k,0)^{\ell}$ such that $\check{Z}'' = (K_*'', U'', \varphi'\rho_2'')$ does not contain a direct summand of type $(0,k,0)$. Moreover, let $\iota'' : U' \longrightarrow F'$ be the inclusion mapping. Then the factorization of ρ over K_*'' gives rise to a factorization $(\rho,0) = (\rho_2'', \iota'')(\rho_1'',0)$ of $(\rho,0)$ over $\check{Z}''$, and $\check{Z}''$ is in $\hat{C}$. Also, $(\rho_1'',0) \in \mathrm{rad}_{\hat{C}}(X,\check{Z}'')$ and $(\rho_2'',\iota'') \in \mathrm{rad}_{\hat{C}}(Z'',X')$, due to the special forms of K, K_*'' and K' . Therefore $(\rho,0) \in \mathrm{rad}^2_{\hat{C}}(X,X')$. q.e.d.

COROLLARY 4.7 (Fourth reduction). If all components of $A(\hat{C})$ are stable tubes then the Auslander-Reiten quivers $A(\hat{S})$ and $A(\hat{C})$ are isomorphic.

PROOF. This is clear from Corollary 4.5 and Lemma 4.6.

5. THE AUSLANDER-REITEN QUIVER OF Λ

Combining the four reductions which have been developed so far, the problem of determining $A_s(\Lambda)$ is now reduced to the problem of determining $A(\hat{C})$ and establishing its tubular structure. In this section we solve this latter problem. Here the main task will be to release the definition of $\hat{C}$ as a full subcategory of $F(K)$ from its technical disguise and to give instead a convenient base-free interpretation of $\hat{C}$, relating it to the module category of a suitable k-algebra A. This algebra A will be a tubular one-point extension of a tame hereditary algebra of extended Dynkin type $\tilde{\mathbb{CD}}_3$, and the solution of our problem emerges from the known Auslander-Reiten structure of modA.

In this context it is convenient to leave the concept of factor-space categories and to switch to the dual concept of subspace categories.

LEMMA 5.1. The categories $\hat{C}^{op}$ and $\hat{D}$ are equivalent.

PROOF. From the definition of $\hat{C}$ and $\hat{D}$ it is immediate that transposition of matrices gives a duality between the categories $\hat{C}$ and $\hat{D}$. q.e.d.

We proceed by defining a particular finite-dimensional k-algebra A which will belong to the class of tubular algebras in the sense of [Ri 84]. The representation theory of such algebras is well-understood. We collect as much information about modA as will be necessary for our purpose. For proofs, and as a general reference, see [Ri 84]. (Note that in [Ri 84] the theory of tubular algebras is developed under the assumption of an algebraically closed base field. However, the generalization to algebras over arbitrary base fields, including the particular algebra A which we are interested in, is straightforward.)

Consider the species

S_0 :

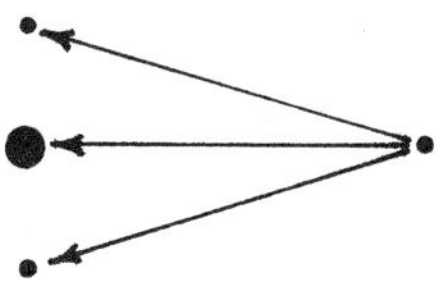

Its path algebra $A_0 = kS_0$ is a tame hereditary algebra of extended Dynkin type $\widetilde{\mathbb{CD}}_3$ and of tubular type $\mathbb{C}_3$. Moreover, consider the representation

R :

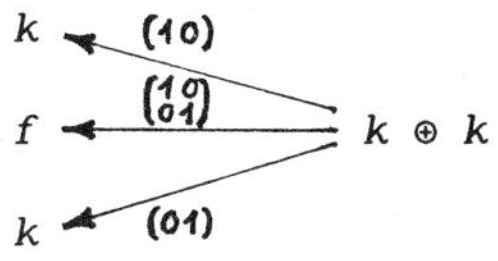

of S_0. Its corresponding A_0-module, also denoted by R, is indecomposable and simple regular.

DEFINITION (The algebra A). We define $A = \begin{pmatrix} A_0 & R \\ 0 & k \end{pmatrix}$ to be the one-point extension of A_0 by R. Alternatively, it can be described as $A = kS/I$, where the species S and a system of relations generating the ideal I are given as follows. (Let $\xi = X + (\delta)$ in $f = k[X]/(\delta)$.)

S:

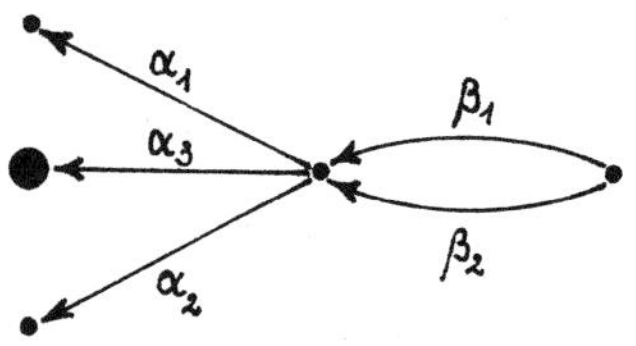

$$\alpha_1\beta_1 = 0$$
$$\alpha_2\beta_2 = 0$$
$$\xi\alpha_3\beta_1 = \alpha_3\beta_2$$
$$d\alpha_3\beta_1 = \xi\alpha_3\beta_2$$

Instead of introducing A as a one-point extension of A_0, we might as well have obtained it by a sequence of three one-point coextensions, starting from $A_\infty = kS_\infty$. where $S_\infty =$ 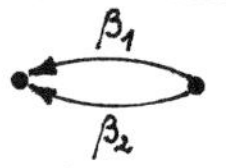

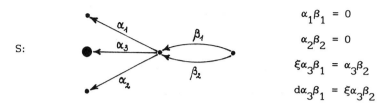

.

Let $\partial_0, \partial_\infty : K_0(A) \longrightarrow \mathbb{Z}$ be the linear forms which are defined by the following coefficients.

$$\partial_0 : \begin{matrix} -1 & & \\ -2 & 2 & 0 \\ -1 & & \end{matrix} \qquad\qquad \partial_\infty : \begin{matrix} 0 & & \\ 0 & -2 & 2 \\ 0 & & \end{matrix}$$

They correspond, up to scalar multiples, to the usual defect forms for the tame hereditary algebras A_0 and A_∞. Moreover, for all $\beta,\alpha \in \mathbb{N}$ let $\partial_{\beta:\alpha} : K_0(A) \longrightarrow \mathbb{Z}$ be the linear form $\partial_{\beta:\alpha} = \beta\partial_\infty + \alpha\partial_0$. For each $\beta:\alpha \in \mathbb{Q}^+_{0,\infty}$ and each $M \in \operatorname{ind}A$ we write briefly $\partial_{\beta:\alpha}(M)$ instead of $\partial_{\beta:\alpha}(\underline{\dim}\, M)$. We denote by S_ω the simple injective A-module, and by S_1, S_2, S_3 the simple projective A-modules, and we set $[S_1,S_2,S_3] = \bigcup_{i=1}^{3}[S_i]$. Then we associate with each $\beta:\alpha \in \mathbb{Q}^+_{0,\infty}$ a partition $\operatorname{ind}A = P_{\beta:\alpha} \,\dot{\cup}\, T_{\beta:\alpha} \,\dot{\cup}\, Q_{\beta:\alpha}$ of indA into three pairwise disjoint classes of indecomposable A-modules, as follows.

$P_0 = \{M \in \operatorname{ind}A \mid \partial_0(M) < 0\}$.

$T_0 = \{M \in \operatorname{ind}A \mid \partial_0(M) = 0\} \setminus [S_\omega]$.

$Q_0 = \{M \in \operatorname{ind}A \mid \partial_0(M) > 0\} \cup [S_\omega]$.

If $\beta:\alpha \in \mathbb{Q}^+$ then $P_{\beta:\alpha}$, respectively $T_{\beta:\alpha}$, respectively $Q_{\beta:\alpha}$ is the set of all indecomposable A-modules M which satisfy $\partial_{\beta:\alpha}(M) < 0$, respectively $\partial_{\beta:\alpha}(M) = 0$, respectively $\partial_{\beta:\alpha}(M) > 0$.

$\mathcal{P}_\infty = \{M \in \text{ind}A \mid \partial_\infty(M) < 0\} \cup [S_1,S_2,S_3]$.

$\mathcal{T}_\infty = \{M \in \text{ind}A \mid \partial_\infty(M) = 0\} \setminus [S_1,S_2,S_3]$.

$\mathcal{Q}_\infty = \{M \in \text{ind}A \mid \partial_\infty(M) > 0\}$.

Saying that $\mathcal{P}_0$, respectively $\mathcal{Q}_\infty$, "is a component of $\mathcal{A}(A)$" , we mean that the set of isomorphism classes of $\mathcal{P}_0$, respectively of $\mathcal{Q}_\infty$, is equal to the set of points of a component of $\mathcal{A}(A)$. Similarly, saying that $\mathcal{T}_{\beta:\alpha}$ "is a tubular I-series of $\mathcal{A}(A)$" we mean that the set of isomorphism classes of $\mathcal{T}_{\beta:\alpha}$ is equal to the set of points of a tubular I-series of $\mathcal{A}(A)$.

Recall that $\mathcal{T}_{\beta:\alpha}$ is said to be a <u>separating tubular I-series, separating $\mathcal{P}_{\beta:\alpha}$ from $\mathcal{Q}_{\beta:\alpha}$</u> , if the following conditions hold:

(s_1) $\mathcal{T}_{\beta:\alpha} = \dot{\bigcup}_{\lambda\in I} \mathcal{T}_{\beta:\alpha}(\lambda)$ is a tubular I-series of $\mathcal{A}(A)$.

(s_2) For all $P \in \mathcal{P}_{\beta:\alpha}$, $T \in \mathcal{T}_{\beta:\alpha}$, $Q \in \mathcal{Q}_{\beta:\alpha}$, the morphism spaces $\text{Hom}_A(Q,P)$, $\text{Hom}_A(Q,T)$, $\text{Hom}_A(T,P)$ are zero.

(s_3) For all $P \in \mathcal{P}_{\beta:\alpha}$, $Q \in \mathcal{Q}_{\beta:\alpha}$, $\varphi \in \text{Hom}_A(P,Q)$, $\lambda \in I$, there exists a factorization of φ through a module in $\text{add}(\mathcal{T}_{\beta:\alpha}(\lambda))$.

We are now prepared to state the main results concerning the structure of $\text{mod}A$.

(A1) For all $\beta:\alpha \in \mathbb{Q}^+_{0,\infty}$, $\mathcal{T}_{\beta:\alpha}$ is a separating tubular I-series, separating $\mathcal{P}_{\beta:\alpha}$ from $\mathcal{Q}_{\beta:\alpha}$.

(A2) $\mathcal{T}_0$ is of left-tubular type $\mathbb{C}_3$ and of right-tubular type $\widetilde{\mathbb{CD}}_3$. For all $\beta:\alpha \in \mathbb{Q}^+$, $\mathcal{T}_{\beta:\alpha}$ is of tubular type $\widetilde{\mathbb{CD}}_3$. $\mathcal{T}_\infty$ is of left-tubular type $\widetilde{\mathbb{CD}}_3$ and of right-tubular type $\mathbb{A}_1$.

(A3) P_0 is a component of $A(\mathrm{A})$. It coincides with the preprojective component in $A(\mathrm{A}_0)$. Q_∞ is a component of $A(\mathrm{A})$. It coincides with the preinjective component in $A(\mathrm{A}_\infty)$. The Auslander-Reiten quiver of A is given by

$$A(\mathrm{A}) = P_0 \dot{\cup} T_0 \dot{\cup} \left(\dot{\bigcup}_{\beta:\alpha\in\mathbb{Q}^+} T_{\beta:\alpha} \right) \dot{\cup} T_\infty \dot{\cup} Q_\infty .$$

On the next page we try to visualize the Auslander-Reiten quiver of A. Each of the separating tubular series $T_{\beta:\alpha}$ is represented by a half line, and the set of all half lines has to be imagined as being dense in the space between the bordering half lines T_0 and T_∞. For each separating tubular series $T_{\beta:\alpha}$, $P_{\beta:\alpha}$ consists of all points in $A(\mathrm{A})$ which are left of $T_{\beta:\alpha}$, and $Q_{\beta:\alpha}$ consists of all points which are right of $T_{\beta:\alpha}$. Consequently, in terms of the picture, all morphisms in modA go from left to right.

Recall that a representation of the bimodule ${}_{\mathrm{A}_0}\mathrm{R}_k$ is given by a triple $(\mathrm{U},\mathrm{M},\varphi)$, where $\mathrm{U} \in \mathrm{mod}k$, $\mathrm{M} \in \mathrm{modA}_0$, and $\varphi \in \mathrm{Hom}_{\mathrm{A}_0}(\mathrm{R}\otimes\mathrm{U},\mathrm{M})$. We denote by $B(\mathrm{R})$ the category of all representations of the bimodule ${}_{\mathrm{A}_0}\mathrm{R}_k$. Let $\mathfrak{Q}$ be the full subcategory of $B(\mathrm{R})$ consisting of all objects $(\mathrm{U},\mathrm{M},\varphi)$ such that M is in $\mathrm{preinjA}_0$. Moreover, let $\mathrm{Hom}_{\mathrm{A}_0}(\mathrm{R},\mathrm{preinjA}_0)$ be the vectorspace category in $\mathrm{mod}k$ which is defined as the image of the functor $\mathrm{Hom}_{\mathrm{A}_0}(\mathrm{R},-) : \mathrm{preinjA}_0 \longrightarrow \mathrm{mod}k$. We denote by $U(\mathrm{Hom}_{\mathrm{A}_0}(\mathrm{R},\mathrm{preinjA}_0))$ its corresponding subspace category. Then, since A is the one-point extension of A_0 by R, there is an equivalence of categories $\Psi_0 : \mathrm{modA} \longrightarrow B(\mathrm{R})$. In addition, adjointness

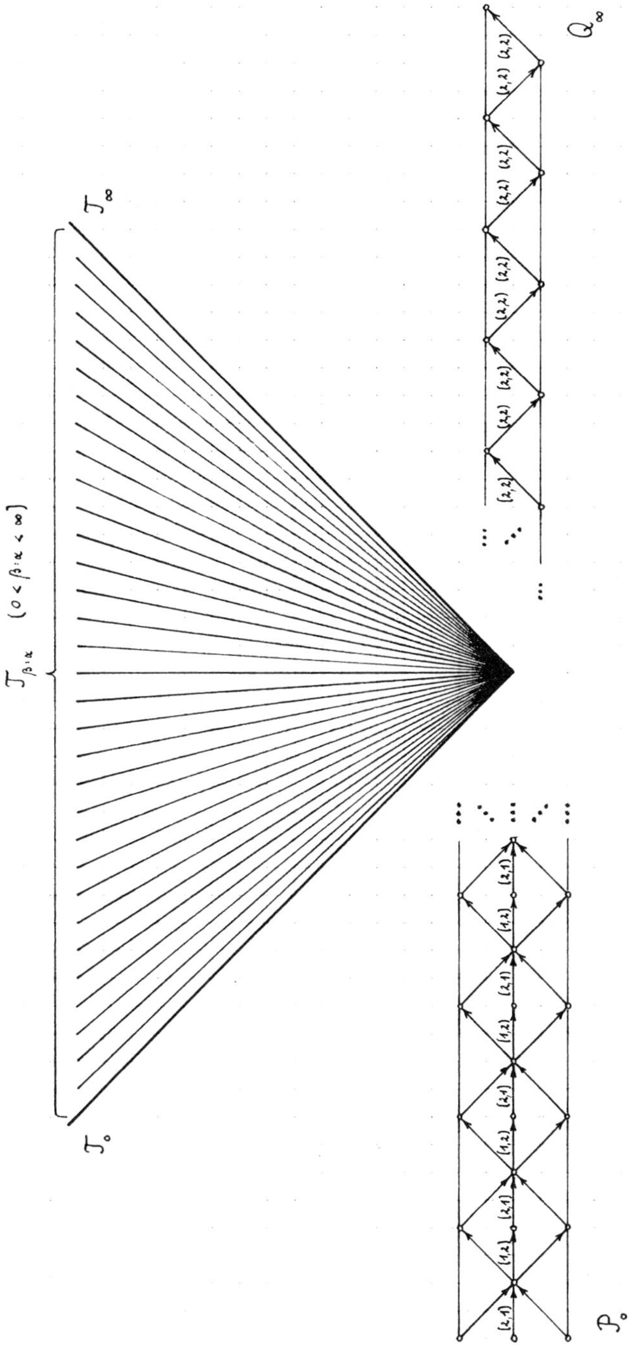

$\mathcal{T}_\infty$
$\mathcal{T}_{\beta:\alpha}$ $(0 < \beta:\alpha < \infty)$
$\mathcal{T}_0$
$\mathcal{Q}_\infty$
$\mathcal{P}_0$

of $R\otimes_k -$ and $\mathrm{Hom}_{A_0}(R,-)$ gives rise to a representation equivalence $\Psi_1 : \mathfrak{Q} \longrightarrow U(\mathrm{Hom}_{A_0}(R,\mathrm{preinj}A_0))$. The sequence of functors $\mathrm{mod}A \xrightarrow{\Psi_0} B(R) \hookleftarrow \mathfrak{Q} \xrightarrow{\Psi_1} U(\mathrm{Hom}_{A_0}(R,\mathrm{preinj}A_0))$ induces a sequence of quiver isomorphisms $A_{\mathrm{mod}A}(Q_0) \cong A(Q_0) \cong A(\mathfrak{Q}) \cong$ $\cong A(U(\mathrm{Hom}_{A_0}(R,\mathrm{preinj}A_0)))$. Therefore we may view $A(U(\mathrm{Hom}_{A_0}(R,\mathrm{preinj}A_0)))$ as being embedded in $A(A)$. Combining these remarks with (A1)-(A3), we obtain the following result concerning the structure of the subspace category $U(\mathrm{Hom}_{A_0}(R,\mathrm{preinj}A_0))$.

(A4) The Auslander-Reiten quiver of $U(\mathrm{Hom}_{A_0}(R,\mathrm{preinj}A_0))$ is given as $A(U(\mathrm{Hom}_{A_0}(R,\mathrm{preinj}A_0))) = \dot{\bigcup}_{\beta:\alpha\in\mathbb{Q}^+} T_{\beta:\alpha} \,\dot{\cup}\, T_\infty \,\dot{\cup}\, Q_\infty$. Moreover, for all $\beta:\alpha \in \mathbb{Q}_\infty^+$, $T_{\beta:\alpha}$ is a separating tubular I-series in $U(\mathrm{Hom}_{A_0}(R,\mathrm{preinj}A_0))$, separating $\dot{\bigcup}_{0<\beta':\alpha'<\beta:\alpha} T_{\beta':\alpha'}$ from $\dot{\bigcup}_{\beta:\alpha<\beta':\alpha'<\infty} T_{\beta':\alpha'} \,\dot{\cup}\, T_\infty \,\dot{\cup}\, Q_\infty$.

We return to our problem of describing the Auslander-Reiten quiver of $\hat{C}$, respectively of $\hat{D}$. The link between this problem and the above short course on modA is provided by the following observation.

LEMMA 5.2. The vectorspace categories $\tilde{K}$ and $\mathrm{Hom}_{A_0}(R,\mathrm{preinj}A_0)$ are equivalent.

PROOF. This amounts to a calculation of the pattern

$\mathrm{Hom}_{A_0}(R,\mathrm{preinj}A_0)$. Since the preinjective component of A_0 is known, this is fairly routine. For a guidance to the calculation of patterns, see [Ri 79].

Henceforth we shall identify the vectorspace categories $\tilde{K}$ and $\mathrm{Hom}_{A_0}(R,\mathrm{preinj}A_0)$, as well as their corresponding subspace categories.

So far we know that $\hat{D}$ is a full subcategory of $U(\tilde{K})$ (by definition), and we have a base-free description of $\tilde{K}$ (by Lemma 5.2) which gives us the knowledge of the Auslander-Reiten quiver of $U(\tilde{K})$ (by (A4)). However, the embedding $\hat{D} \hookrightarrow U(\tilde{K})$ is described in terms of matrices. Therefore our next task is to release the definition of $\hat{D}$ as a full subcategory of $U(\tilde{K})$ from its technical diguise and to give instead an equivalent description in base-free terms. Moreover, it will turn out that this new description is adapted to the Auslander-Reiten structure of $U(\tilde{K})$ so well that the desired Auslander-Reiten quiver of $\hat{D}$ can be easily determined as a full subquiver of the Auslander-Reiten quiver of $U(\tilde{K})$.

We fix some terminology and notation. In general, objects in $U(\tilde{K})$ will be denoted by $X = (U,K,\psi)$, with $K = \bigoplus_{i=-\infty}^{8} K_i^{n_i}$. For any object K in $\tilde{K}$, the <u>support</u> of K is the set of all indecomposable direct summands of K . For objects $X = (U,K,\psi)$ in $U(\tilde{K})$ we define support(X) = support(K). Recall that for any pair of integers (a,b) , subject to $a \leq b \leq 8$, we denote by $\tilde{K}_{[a,b]}$ the full subcategory $\mathrm{add}\{K_i \mid i \in [a,b]\}$ of $\tilde{K}$. If $(d_1),\ldots,(d_n)$ is a set of well-defined conditions for the objects in $U(\tilde{K})$ then we denote by

$U(\tilde{K})_{(d_1,\dots,d_n)}$ the full subcategory of $U(\tilde{K})$ consisting of all objects which satisfy each of the conditions $(d_1),\dots,(d_n)$.

We introduce successively conditions $(d_1),\dots,(d_5)$ for the objects X in $U(\tilde{K})$. We begin with the formulation of condition (d_1).

(d_1) : $\mathrm{support}(X) \subset \{K_i \mid -3 \le i \le 8\}$.

Let $X = (U,K,\psi)$, with $K = \bigoplus_{i=-3}^{8} K_i^{n_i}$, be an object in $U(\tilde{K})_{(d_1)}$, and denote by $\alpha_{-1} : K \twoheadrightarrow \bigoplus_{i=-3}^{8} K_i^{n_i}$ the projection. Then the assignment $X \longmapsto A_{-1}(X) = (U, \bigoplus_{i=-3}^{-1} K_i^{n_i} , \alpha_{-1}\psi)$ defines a functor $A_{-1} : U(\tilde{K})_{(d_1)} \to U(\tilde{K}_{[-3,-1]}) \cong \mathrm{mod}kS_o$. Note that, for all $X \in U(\tilde{K})_{(d_1)}$, $A_{-1}(X) \cong X/(0, \bigoplus_{i=0}^{8} K_i^{n_i} , 0)$ is a factorobject of X . Denote by I_i , $i=4,\dots,8$, the indecomposable preinjective objects in $\mathrm{mod}kS_o$ which represent the points in the last mesh of $A(S_o)$. To be precise, we indicate this mesh together with the dimension types of I_i (viewed as objects in $\mathrm{mod}kS_o$) , and together with chosen normal forms for the I_i (viewed as objects in $U(\tilde{K}_{[-3,0]})$).

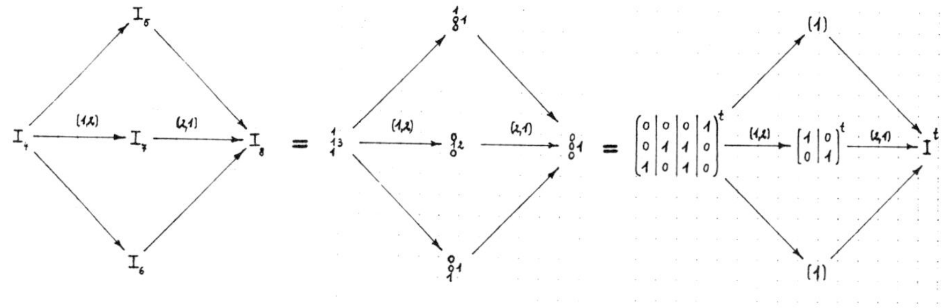

Then we can formulate condition (d_2).

(d_2): $\quad X \in \mathcal{U}(\tilde{K})_{(d_1)}$ and $A_{-1}(X) \in \operatorname{add}\{I_i \mid i=4,\ldots,8\}$.

Let $X = (U,K,\psi)$, with $K = \bigoplus_{i=-3}^{8} K_i^{n_i}$, be an object in $\mathcal{U}(\tilde{K})_{(d_1)}$, and denote by $\alpha_o : K \twoheadrightarrow \bigoplus_{i=-3}^{0} K_i^{n_i}$ the projection. Then the assignment $X \longmapsto A_o(X) = (U, \bigoplus_{i=-3}^{0} K_i^{n_i}, \alpha_o\psi)$ defines a functor $A_o : \mathcal{U}(\tilde{K})_{(d_1)} \longrightarrow \mathcal{U}(\tilde{K}_{[-3,0]})$. Note that, for all objects $X \in \mathcal{U}(\tilde{K})_{(d_1)}$, $A_o(X) \cong X/(0 , \bigoplus_{i=1}^{8} K_i^{n_i} , 0)$ is a factorobject of X . We introduce full subcategories $\mathcal{Y}$, $\overline{\mathcal{Y}}$, $\mathcal{Z}$ of $\mathcal{U}(\tilde{K}_{[-3,0]}$. Let $\mathcal{Y}$ consist of all objects $\vartheta(X,Y,Z)$, where $(X,Y,Z) \in \mathcal{A}$ satisfies the column-conditions $n_i = 0$ for all $i \in I_X\backslash\{0\}$. Let $\overline{\mathcal{Y}}$ be the closure of $\mathcal{Y}$ in $\mathcal{U}(\tilde{K}_{[-3,0]})$ under isomorphism classes. Let $\mathcal{Z}$ consist of all objects $\vartheta(X,Y,Z)$, where $(X,Y,Z) \in \mathcal{A}$ satisfies the column conditions $n_i = 0$ for all $i \in I_X\backslash\{0\}$ and the row conditions $m_j = 0$ for all $j \in I_Y\backslash\{4\}$. By definition of (d_2) and $\overline{\mathcal{Y}}$, the restriction of A_o to $\mathcal{U}(\tilde{K})_{(d_1,d_2)}$ yields a functor $A_o : \mathcal{U}(\tilde{K})_{(d_1,d_2)} \longrightarrow \overline{\mathcal{Y}}$. Let $A_1 : \mathcal{Y} \hookrightarrow \overline{\mathcal{Y}}$ be the inclusion functor. Being full, faithful and dense, it is an equivalence of categories. Let $Y = (U , \bigoplus_{i=-3}^{0} K_i^{n_i} , \upsilon)$ be an object in $\mathcal{Y}$. Then $A_{-1}(Y) = \bigoplus_{i=0}^{4} I_i^{m_i}$, where $m_{i+4} + m_4 = n_i$ for all $i = -3,-2,-1$. Denote by $\pi_o : \bigoplus_{i=-3}^{0} K_i^{n_i} \twoheadrightarrow K_o^{n_o}$ the projection. Then there is a unique subobject $Y' = (U' , \bigoplus_{i=-3}^{0} K_i^{n'_i} , \upsilon\iota)$ of Y, where $\iota : U' \hookrightarrow U$ is the inclusion, such that $A_{-1}(Y') = \bigoplus_{i=0}^{3} I_i^{m_i}$ and

$\pi_0 \upsilon\iota : U' \longrightarrow K_0^{n_0'}$, viewed as a Kronecker-module, has no simple projective direct summand. Hence $m_{i+4} = n_i'$ for all $i = -3,-2,-1$. The assignment $Y \longmapsto A_2(Y) = Y/Y' = (U/U' , \bigoplus_{i=-3}^{0} K_i^{n_i - n_i'} , \upsilon/\upsilon\iota)$ defines a functor $A_2 : \mathcal{Y} \longrightarrow \mathcal{Z}$. Let $Z = (U,K,\psi)$ be an object in $\mathcal{Z}$, given by the matrix $\vartheta(X,Y,Z)$. Then the assignment $Z \longmapsto A_3(Z) = k^{n} \underset{\eta'}{\overset{\eta}{\leftleftarrows}} k^{m_4}$, where the k-linear maps η, η' are given by the matrices $Y_4^t, Y_{4'}^t$ modulo π , defines a functor $A_3 : \mathcal{Z} \longrightarrow \mathrm{mod}kQ_2$ which is easily seen to be an equivalence of categories. Altogether we have introduced the following sequence of categories and functors.

$$\begin{array}{ccccccccc}
\mathcal{U}(\tilde{K})_{(d_1)} & \longrightarrow & \mathcal{U}(\tilde{K}_{[-3,0]}) & & & & & & \\
\uparrow & & \uparrow & & & & & & \\
\mathcal{U}(\tilde{K})_{(d_1,d_2)} & \xrightarrow{A_o} & \bar{\mathcal{Y}} & \xleftarrow{A_1} & \mathcal{Y} & \xrightarrow{A_2} & \mathcal{Z} & \xrightarrow{A_3} & \mathrm{mod}kQ_2
\end{array}$$

Let $A : \mathcal{U}(\tilde{K})_{(d_1,d_2)} \longrightarrow \mathrm{mod}kQ_2$ be the compositions of functors $A = A_3 A_2 A_1^{-1} A_o$, where A_1^{-1} is a chosen inverse for the equivalence A_1 . Note that, for all $X \in \mathcal{U}(\tilde{K})_{(d_1,d_2)}$, $A(X)$ is a factorobject of X . With this functor A we can now formulate condition (d_3) .

(d_3) : $\qquad X \in \mathcal{U}(K)_{(d_1,d_2)}$ and $A(X) \in \mathrm{preinj}kQ_2$.

Let $X = (U,K,\psi)$, with $K = \bigoplus_{i=-3}^{8} K_i^{n_i}$, be an object in $\mathcal{U}(\tilde{K})_{(d_1)}$, and denote by $\alpha_7 : K \longrightarrow\!\!\!\!\rightarrow \bigoplus_{i=-3}^{7} K_i^{n_i}$ the projection. Let $U' = \ker\alpha_7\psi$ and $\iota : U' \hookrightarrow U$ the inclusion. Then the assignment $X \longmapsto \Omega(X) = (U', K_8^{n_8}, \psi\iota)$ defines a functor $\Omega : \mathcal{U}(\tilde{K})_{(d_1)} \longrightarrow \mathrm{mod}kQ_2$.

Note that, for all $X \in U(\tilde{K})_{(d_1)}$, $\Omega(X)$ is a subobject of X. With this functor Ω we formulate condition (d_4).

(d_4) : $X \in U(\tilde{K})_{(d_1)}$ and $\Omega(X) \in \text{preproj}\, kQ_2 \vee (\bigvee_{\lambda \in \check{I}} \text{reg}_\lambda kQ_2)$.

Consider the following six particular objects in $U(\tilde{K})$.

$X_{\alpha,\bar{\infty}} = (k, K_{-2} \oplus K_o, (1\|0|1)^t)$

$X_{\alpha,o} = (k, K_{-3} \oplus K_o, (1\|1|0)^t)$

$X_{\alpha,\delta} = (k^2, K_{-1} \oplus K_0^2, \left(\begin{smallmatrix}1&0\\0&1\end{smallmatrix}\middle\|\begin{smallmatrix}1&0\\0&1\end{smallmatrix}\middle|\begin{smallmatrix}0&0\\0&0\end{smallmatrix}\right)^t)$

$X_{\omega,\bar{\infty}} = (k, K_5, (1))$

$X_{\omega,\bar{o}} = (k, K_6, (1))$

$X_{\omega,\bar{\delta}} = (k^2, K_7, \left(\begin{smallmatrix}1&0\\0&1\end{smallmatrix}\right)^t)$

Then we can formulate condition (d_5).

(d_5) : No direct summand of X is isomorphic to an object in

$\{X_{\alpha,\bar{\infty}}, X_{\alpha,o}, X_{\alpha,\delta}, X_{\omega,\bar{\infty}}, X_{\omega,\bar{o}}, X_{\omega,\bar{\delta}}\}$.

We are interested in the full subcategory $U(\tilde{K})_{(d_1,\ldots,d_5)}$ of $U(\tilde{K})$ which we henceforth abbreviate by $U(\tilde{K})_{(d)}$.

LEMMA 5.3. The categories $\hat{D}$ and $U(\tilde{K})_{(d)}$ are equivalent.

PROOF. For any angular matrix $(X,Y,Z) \in \mathcal{A}$ the following statements hold.

(1) If $\lambda(X,Y,Z) \in \text{preproj}\, kQ_3$ then $A(\vartheta(X,Y,Z)) \in \text{preinj}\, kQ_2$.

(2) If $\rho(X,Y,Z) \in \text{preinj}kQ_3 \vee (\bigvee_{\lambda\in\check{I}} \text{reg}_\lambda kQ_3)$ then

$\Omega(\vartheta(X,Y,Z)) \in \text{preproj}kQ \vee (\bigvee_{\lambda\in\check{I}} \text{reg}_\lambda kQ_2)$.

(3) If $A(\vartheta(X,Y,Z)) \in \text{preinj}kQ_2$ then $\lambda(X,Y,Z) \in \text{preproj}kQ_3 \vee \{R(\bar{\infty})_1\}$.

(4) If $R(\bar{\infty})_1$ is a direct summand of $\lambda(X,Y,Z)$ then $X_{\alpha,\bar{\infty}}$ is a direct summand of $\vartheta(X,Y,Z)$.

(5) If $\Omega(\vartheta(X,Y,Z)) \in \text{preproj}kQ_2 \vee (\bigvee_{\lambda\in\check{I}} \text{reg}_\lambda kQ_2)$ then $\rho(X,Y,Z) \in \text{preinj}kQ_3 \vee (\bigvee_{\lambda\in\check{I}} \text{reg}_\lambda kQ_3) \vee \{R(\bar{\infty})_1\}$.

(6) If $R(\bar{\infty})_1$ is a direct summand of $\rho(X,Y,Z)$ then $X_{\omega,\bar{\infty}}$ is a direct summand of $\vartheta(X,Y,Z)$.

Indeed, recurring to the definition of $\lambda,\rho,\vartheta,A,\Omega$ and using the normal forms for $\text{ind}kQ_2$ and $\text{ind}kQ_3$, as presented in 0.3, the verification of (1)-(6) is straightforward.

Let $X = (U,K,\psi) = \vartheta(X,Y,Z)$, with $(X,Y,Z) \in \mathfrak{M}$, be an object in $\hat{D}$. Then X satisfies conditions (d_1) and (d_2), by construction of ϑ. Moreover, X satisfies conditions (d_3) and (d_4) , by definition of $\mathfrak{M}$ in combination with (1) and (2). Finally, X satisfies condition (d_5), by definition of $\mathfrak{M}$. Hence X is an object in $U(K)_{(d)}$.

Conversely, let X be an object in $U(K)_{(d)}$. Because X satisfies (d_1) and (d_2), there exists an angular matrix $(X,Y,Z) \in \mathcal{A}$ such that $X \cong \vartheta(X,Y,Z)$. Moreover, X satisfies (d_3) - (d_5) and therefore $\vartheta(X,Y,Z)$ does so. In view of (3) - (6) this implies that (X,Y,Z) is in $\mathfrak{M}$. Hence $\vartheta(X,Y,Z)$ is in $\hat{D}$.

Thus we have proved that $\hat{D}$ is a dense subcategory of $U(\tilde{K})_{(d)}$. Since both categories $\hat{D}$ and $U(\tilde{K})_{(d)}$ are full subcategories of $U(\tilde{K})$, it follows that they are equivalent. q.e.d.

The definition of $U(\tilde{K})_{(d)}$, as given above in terms of the conditions (d_1) - (d_5), seems to be still rather involved. Indeed, we only need it as an intermediate step towards a much easier comprehensible description of $U(\tilde{K})_{(d)}$, to be given below in terms of the Auslander-Reiten quiver of $U(\tilde{K})$.

Recall from the above discussion of modA that there is a representation equivalence $\Psi_1\Psi_0 : Q_0 \longrightarrow U(\tilde{K})$ which induces a quiver-embedding $A(U(\tilde{K})) \cong A_{\mathrm{modA}}(Q_0) \subset A(\mathrm{A})$. In view of this fact we henceforth denote by $T_{\beta:\alpha}$ not only the class of indecomposable A-modules defined above, but also the class of indecomposable objects in $U(\tilde{K})$ corresponding to them under Ψ_1, Ψ_0, for all $\beta:\alpha \in \mathbb{Q}^+_\infty$. In the sequel we denote indecomposable objects in modA by M, indecomposable objects in $U(\tilde{K})$ by X, and we write $\mathrm{M} = \mathrm{M}(X)$, or else $X = X(\mathrm{M})$, if we want to emphasize their correspondence under $\Psi_1\Psi_0$.

Recall from [Ri84] the notion of the <u>index</u> $\gamma(\mathrm{x})$ of an element $\mathrm{x} \in \mathrm{K}_0(\mathrm{A})$. If $(\partial_0(\mathrm{x}), \partial_\infty(\mathrm{x})) \neq (0,0)$ then we say that the index of x is defined and we put $\gamma(\mathrm{x}) = \dfrac{\partial_0(\mathrm{x})}{-\partial_\infty(\mathrm{x})}$. Hence $\gamma(\mathrm{x})$ is an element of $\mathbb{Q} \cup \{\infty\}$. For all objects $\mathrm{M} \in \mathrm{indA}$ we introduce the index of M on setting $\gamma(\mathrm{M}) = \gamma(\underline{\dim}_{\mathrm{A}}\mathrm{M})$. Also, for all objects $X \in \mathrm{ind}U(\tilde{K})$ we introduce the index of X on setting $\gamma(X) = \gamma(\underline{\dim}_{\mathrm{A}}\mathrm{M}(X))$. In fact, for all $\mathrm{M} \in \mathrm{indA}$ and hence also for all $X \in \mathrm{ind}U(\tilde{K})$, the index of M and the index of X is defined. Moreover, for any $\mathrm{M} \in T_0 \dot{\cup} (Q_0 \cap P_\infty) \dot{\cup} T_\infty$ and any $\beta:\alpha \in \mathbb{Q}^+_{0,\infty}$, $\gamma(\mathrm{M}) = \beta:\alpha$ if and only if $\partial_{\beta:\alpha}(\mathrm{M}) = 0$.

We need some more notation for particular classes of objects in $U(\tilde{K})$. We set

$$U(\tilde{K})_{]\frac{3}{14},\infty[} = \dot{\bigcup}_{\frac{3}{14}<\beta:\alpha<\infty} T_{\beta:\alpha} ,$$

$$U(\tilde{K})_{[\frac{3}{14},\infty]} = \dot{\bigcup}_{\frac{3}{14}\leq\beta:\alpha\leq\infty} T_{\beta:\alpha} ,$$

$$\check{T}_{\infty} = \dot{\bigcup}_{\lambda\in\check{I}} T_{\infty}(\lambda) .$$

PROPOSITION 5.4. The categories $\mathrm{ind}U(\tilde{K})_{(d)}$ and $U(\tilde{K})_{]\frac{3}{14},\infty[} \dot{\cup} \check{T}_{\infty}$ are equal.

PROOF. Repeatedly we shall have to deal with the preinjective component of $A(kS_0)$. We label its points according to the following picture.

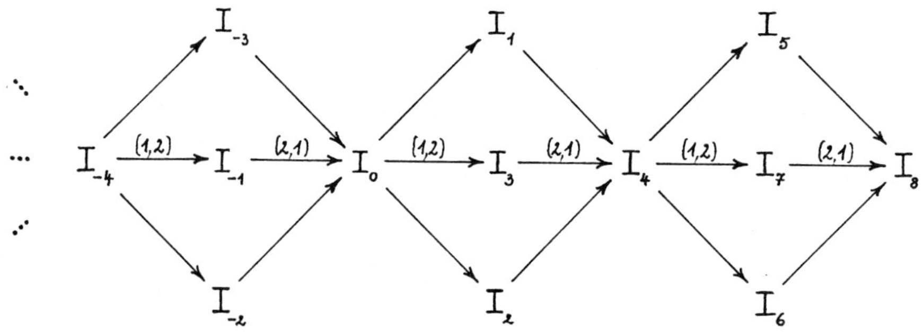

Their dimension types are given by

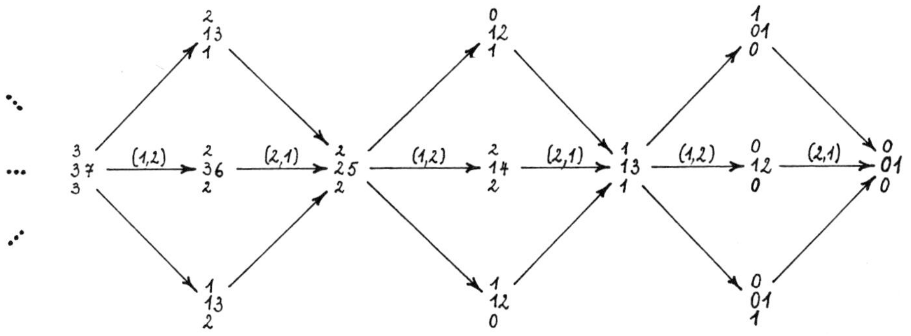

We subdivide the proof into four parts.

(1) We claim that $U(\tilde{K})_{]\frac{3}{14},\infty[} \subset \mathrm{ind}U(\tilde{K})_{(d)}$.

Let $X = (U, K, \psi) \in U(\tilde{K})_{]\frac{3}{14},\infty[}$. We have to show that X satisfies the conditions $(d_1),\ldots,(d_5)$.

The projective cover of $M(X)$ factorizes over a module T in $\mathrm{add}T_{3:14}(\lambda)$, for any chosen $\lambda \in \check{I}$. Hence there is an epimorphism $\rho : T \twoheadrightarrow M(X)$ which induces an epimorphism of A_0-modules $\rho_0 : T|_{A_0} \twoheadrightarrow M(X)|_{A_0}$. It is not difficult to see that $M(\vartheta(X,Y,Z))$, with $(X,Y,Z) = \{Y_4, Y_{4'}\} = \{E, \phi_\lambda\}$ gives the primitive series of

$T_{3:14}$, as λ ranges through I . Therefore $T|_{A_o} = \bigoplus_{i=-3}^{o} I_i^{n}$ and $M(X)|_{A_o} = \bigoplus_{i=-3}^{8} I_i^{n_i}$, which means that X satisfies condition (d_1) .

Consider the factorobject $A_{-1}(X)$ of X . Recall that it can be viewed as an object in $\mathrm{mod}kS_o$. Assume that there is an indecomposable direct summand N of $A_{-1}(X)$ which is not isomorphic to any of the modules $I_4,\ldots,I_8$. Then for some $j \in \{1,2,3\}$ we have $\mathrm{Hom}_{kS_o}(N,I_j) \neq 0$, and consequently $U(\tilde{K})(X,I_j) \neq 0$. On the other hand, $I_j \in T_{3:14}$ for all $j = 1,2,3$, and therefore $U(\tilde{K})(X,I_j) = 0$. This contradiction implies that X satisfies condition (d_2).

Because X is in $U(\tilde{K})_{(d_1,d_2)}$ we have $X \cong \vartheta(X,Y,Z)$, for some (X,Y,Z) in $\mathcal{A}$. Assume that there is an indecomposable direct summand N of $A(X)$ which is not in $\mathrm{preinj}kQ_2$. Then the object $A_3^{-1}(N)$ in $U(\tilde{K})$ is a direct summand of $A_2A_1^{-1}A_o(X)$, which in turn is a factorobject of X . Hence $U(\tilde{K})(X,A_3^{-1}(N)) \neq 0$. On the other hand, $\partial_{3:14}(A_3^{-1}(N)) \leq 0$, and therefore $U(\tilde{K})(X,A_3^{-1}(N)) = 0$. This contradiction shows that X satisfies condition (d_3).

Assume that there is an indecomposable direct summand N of $\Omega(X)$ which is not in $\mathrm{preproj}kQ_2$. Viewing N as subobject of X we see that $U(\tilde{K})(N,X) \neq 0$. On the other hand we find that $\partial_\infty(N) \geq 0$, and therefore $U(\tilde{K})(N,X) = 0$. This contradicion shows that X satisfies condition (d_4).

Our given object X clearly satisfies condition (d_5) , because $\{X_{\alpha,\bar{\infty}}, X_{\alpha,o}, X_{\alpha,\delta}\} \subset T_{3:14}$ and $\{X_{\omega,\bar{\infty}}, X_{\omega,\bar{o}}, X_{\omega,\bar{\delta}}\} \subset T_\infty$.

(2) We claim that $\operatorname{ind}U(\tilde{K})_{(d)} \subset U(\tilde{K})_{[\frac{3}{14},\infty]}$.

Let $X = (U,K,\psi) \in \operatorname{ind}U(\tilde{K})_{(d)}$. Because X satisfies condition (d_4), $\Omega(X)$ is not preinjective, and therefore X is not in Q_∞ . Hence it remains to show that $\gamma(X) \geq \frac{3}{14}$.

Assume that $\gamma(X) < \frac{3}{14}$. Then the injective hull of $M(X)$ factorizes over a module T in $\operatorname{add}T_{3:14}(\lambda)$, for any chosen $\lambda \in \check{I}$. Hence there is a monomorphism $\kappa : M(X) \hookrightarrow T$, which induces a monomorphism of A_o-modules $\kappa_o : M(X)|_{A_o} \hookrightarrow T|_{A_o}$. Looking at the primitive series of $T_{3:14}$, as above, we see that $T|_{A_o} = \bigoplus_{i=-3}^{o} I_i^n$. Because X satisfies condition (d_1) , we conclude that $X = (U, \bigoplus_{i=-3}^{o} K_i^{n_i}, \psi)$. Consider $\tilde{\kappa} = \Psi_1\Psi_o(\kappa) : X \hookrightarrow X(T)$, with $X(T) = (U', \bigoplus_{i=-3}^{o} K_i^n, \psi')$, and consider $A_{-1}(\tilde{\kappa}) : A_{-1}(X) \to A_{-1}(X(T))$. This latter morphism is given by $A_{-1}(\tilde{\kappa}) = (\tilde{\kappa}_u, \tilde{\kappa}_i)_{i=-3,-2,-1}$, where $\tilde{\kappa}_u : U \hookrightarrow U'$ and $\tilde{\kappa}_i : K_i^{n_i} \longrightarrow K_i^n$, for all $i = -3,-2,-1$. Note that $\tilde{\kappa}_u$ is a monomorphism, that $A_{-1}(X) = \bigoplus_{i=4}^{8} I_i^{n_i}$ by condition (d_2) , and that $A_{-1}(X(T)) = I_4^n$. Altogether this yields that $A_{-1}(X) = I_4^{n_4}$. Therefore $X = \vartheta(X,Y,Z)$, for an angular matrix $(X,Y,Z) = \{Y_4, Y_{4'}\}$. Since X satisfies condition (d_3) , $A(X)$ is in $\operatorname{preinj}kQ_2$. It follows that $\gamma(X) > \frac{3}{14}$, contradicting our assumption. Thus we have proved assertion (2).

(3) We claim that $T_{3:14} \cap \operatorname{ind}U(\tilde{K})_{(d)} = \emptyset$.

Assume that there exists an object $X = (U,K,\psi)$ in $T_{3:14} \cap \mathrm{ind}U(\tilde{K})_{(d)}$. Because $\hat{D}$ is a dense subcategory of $U(\tilde{K})_{(d)}$ (see proof of Lemma 5.3), we may assume that $X = \vartheta(X,Y,Z)$, where $(X,Y,Z) \in \mathfrak{M}$. As usual let $\{m_j,n_j\} \subset \mathbb{N}_o$ be the set of natural numbers indicating the sizes of the submatrices X_i,Y_j,Z_k of (X,Y,Z).

Suppose that there exists an $i \in \{1,\dots,8\}$ such that $K_i \in \mathrm{support}(X)$. Then $X_1 = (0,K_i,o)$ is a subobject of X, and therefore $U(\tilde{K})(X_1,X) \neq 0$. On the other hand, $\gamma(X_1) > \frac{3}{14}$, and therefore $U(\tilde{K})(X_1,X) = 0$. This contradiction shows that $\mathrm{support}(X) \subset \subset \{K_{-3},\dots,K_o\}$.

Consider the subobject $X_2 = (U_2,K_o^{n_o},\psi_2)$ of X, which is given by $U_2 = \ker(\alpha_{-1}\psi)$, $\psi_2 = \psi|_{U_2}$, and $\alpha_{-1} : K \twoheadrightarrow \bigoplus_{i=-3}^{-1} K_i^{n_i}$ the projection. Let N_2 be a nonzero indecomposable direct summand of X_2. Then $U(\tilde{K})(N_2,X) \neq 0$. On the other hand, $\gamma(N_2) > \frac{3}{14}$ and therefore $U(\tilde{K})(N_2,X) = 0$ whenever $N_2 \neq (0,K_o,o)$. Hence $X_2 = (0,K_o^{n_o},o)$, which implies that $m_o = 0$.

So far we know that $X = \vartheta(X,Y,Z)$, where $(X,Y,Z) \in \mathfrak{M}$, subject to $n_i = 0$ for all $i \in I_X\backslash\{0\}$, and $m_o = 0$. Now consider the subobject $X_3 = (U_3, \bigoplus_{i=-3}^{-1} K_i^{m_i} \oplus K_o^{n_o},\psi_3)$ of X which is uniquely determined by the conditions $A_{-1}(X_3) = \bigoplus_{i=5}^{7} I_i^{m_i}$ and $A(X_3)$ does not contain a simple projective direct summand. Note that $U_3 = k^{m_1+m_2+2m_3}$. Let N_3 be a nonzero indecomposable direct summand of X_3. Then $U(\tilde{K})(N_3,X) \neq 0$. On the other hand, $\gamma(N_3) > \frac{3}{14}$ and therefore $U(\tilde{K})(N_3,X) = 0$ whenever N_3 is not isomorphic to an object in $\{X_{\alpha,o}, X_{\alpha,\bar{\infty}}, X_{\alpha,\delta}\}$. Hence $X_3 \cong X_{\alpha,o}^{m_1} \oplus X_{\alpha,\bar{\infty}}^{m_2} \oplus X_{\alpha,\delta}^{m_3}$.

Set $n''_o = n_o - n'_o$. Then $\underline{\dim}_A M(X)$ is given by

$(4m_1+3m_2+6m_3+5m_4+2n''_o)$

$(3m_1+3m_2+7m_3+5m_4+2n''_o)$ $(8m_1+8m_2+16m_3+12m_4+5n''_o)$ $(m_1+m_2+2m_3+3m_4)$

$(3m_1+4m_2+6m_3+5m_4+2n''_o)$

An easy calculation shows that the equation $\gamma(X) = \gamma(\underline{\dim}_A M(X)) = \frac{3}{14}$, which is true by assumption, implies that $n''_o = m_4$. If $A(X) \neq 0$ then $n''_o = m_4$ contradicts condition (d_3) . If $A(X) = 0$ then $X = X_3$, which contradicts condition (d_5) . This proves assertion (3).

(4) We claim that $T_\infty \cap \operatorname{ind}U(\tilde{K})_{(d)} = \check{T}_\infty$.

Observe that $\check{T}_\infty$ consists of all indecomposable objects X in $U(\tilde{K})$ for which $X = \Omega(X) \in \bigvee_{\lambda\in \check{I}} \operatorname{reg}_\lambda kQ_2$. Hence it is clear that $\check{T}_\infty \subset T_\infty \cap \operatorname{ind}U(\tilde{K})_{(d)}$.

Conversely, let $X = (U,K,\psi)$ be in $T_\infty \cap \operatorname{ind}U(\tilde{K})_{(d)}$. As in the proof of (3), we may assume that $X = \vartheta(X,Y,Z)$, where $(X,Y,Z) \in \mathfrak{M}$. Consider the subobject $X_1 = (U_1, \bigoplus_{i=0}^{8} K_i^{n_i}, \psi_1)$ of X which is given by $U_1 = \ker(\alpha_{-1}\psi)$, $\psi_1 = \psi|_{U_1}$, and let $X/X_1 = (U/U_1, \bigoplus_{i=-3}^{-1} K_i^{n_i}, \psi/\psi_1)$ be the corresponding factorobject. Let N_1 be a nonzero indecomposable direct summand of X/X_1 . Then $U(\tilde{K})(X,N_1) \neq 0$. On the other hand, $\gamma(N_1) < \infty$, and therefore $U(\tilde{K})(X,N_1) = 0$. Hence $X/X_1 = 0$, in other words $X = X_1$.

Proceeding along this line and looking successively at

factorobjects of X with support $\{K_o\}$, respectively $\{K_1,K_2,K_3\}$, respectively $\{K_4\}$, respectively $\{K_5,K_6,K_7\}$, we eventually find that $X = (U, \bigoplus_{i=5}^{8} K_i^{n_i}, \psi)$ and $X/(0,K_8^{n_8},o) \cong X_{\omega,\infty}^{n_5} \oplus X_{\omega,o}^{n_6} \oplus X_{\omega,\delta}^{n_7} \oplus (k,0,o)^{m_o'}$.

Then $\underline{\dim}_A M(X)$ is given by

$$\begin{matrix} n_5 & & \\ n_7 & (n_5+n_6+2n_7+n_8) & (m_o'+n_5+n_6+2n_7) \\ n_6 & & \end{matrix} .$$

The equation $\partial_\infty(\underline{\dim}_A M(X)) = 0$, which is valid by assumption, implies that $m_o' = n_8$. Because X satisfies condition (d_4) , it follows that $\Omega(X) \in \bigvee_{\lambda\in \check{I}} \mathrm{reg}_\lambda kQ_2$ and that $X = X_{\omega,\infty}^{n_5} \oplus X_{\omega,o}^{n_6} \oplus X_{\omega,\delta}^{n_7} \oplus \Omega(X)$. However, X satisfies condition (d_5) and is indecomposable. Therefore $X \in \check{T}_\infty$. This proves assertion (4).

By (1) - (4) it is clear that $\mathrm{ind}U(\tilde{K})_{(d)} = U(\tilde{K})_{]\frac{3}{14},\infty[} \,\dot\cup\, \check{T}_\infty$.

q.e.d.

We set $\tilde{T}_\infty = \dot\bigcup_{\lambda\in I} \tilde{T}_\infty(\lambda)$, where $\begin{cases} \tilde{T}_\infty(\lambda) = T_\infty(\lambda) & , \text{ if } \lambda \in I\setminus E \\ \tilde{T}_\infty(\lambda) = A_{\mathcal{M}}(T(\lambda))^{op} & , \text{ if } \lambda \in E \end{cases}$.

Note that $\tilde{T}_\infty$ is a tubular I-series of tubular type $\widetilde{\mathbb{CD}}_3$. We are now in the position to describe $A(\hat{C})$, as well as $A_s(\Lambda)$.

COROLLARY 5.5. There are the following isomorphisms of quivers:

(i) $A(\hat{C})^{op} \cong \dot\bigcup_{\frac{3}{14}<\beta:\alpha<\infty} T_{\beta:\alpha} \,\dot\cup\, \tilde{T}_\infty$.

(ii) $A_s(\Lambda)^{op} \cong \dot\bigcup_{\frac{3}{14}<\beta:\alpha<\infty} T_{\beta:\alpha} \,\dot\cup\, \tilde{T}_\infty$.

PROOF. (i) Combining Lemma 5.1, Lemma 5.3, Proposition 5.4, the separating property of $T_{3:14}$ and of T_∞, and our knowledge of $A(U(\tilde{K}))$, we obtain the following sequence of quiver isomorphisms:

$$A(\hat{C})^{op} \cong A(\hat{D}) \cong A(U(\tilde{K})_{(d)}) \cong A(\mathrm{add}(U(\tilde{K})_{]\frac{3}{14},\infty[} \;\dot{\cup}\; \check{T}_\infty)) \cong$$

$$\cong A_{U(\tilde{K})}(U(\tilde{K})_{]\frac{3}{14},\infty[} \;\dot{\cup}\; \check{T}_\infty) \cong \dot{\bigcup}_{\frac{3}{14}<\beta:\alpha<\infty} T_{\beta:\alpha} \;\dot{\cup}\; \check{T}_\infty .$$

(ii) Note that (i) implies in particular that all components of $A(\hat{C})$ are stable tubes. Combining Corollaries 1.3, 2.2, 3.6, 4.7 and (i), we obtain the following sequence of quiver isomorphisms:

$$A_s(\Lambda)^{op} \cong A(F(K_1))^{op} \cong A_{\mathcal{M}}(\hat{S} \vee T(E))^{op} \cong A(\hat{S})^{op} \;\dot{\cup}\; A_{\mathcal{M}}(T(E))^{op} \cong$$

$$\cong A(\hat{C})^{op} \;\dot{\cup}\; A_{\mathcal{M}}(T(E))^{op} \cong \dot{\bigcup}_{\frac{3}{14}<\beta:\alpha<\infty} T_{\beta:\alpha} \;\dot{\cup}\; \tilde{T}_\infty .$$

q.e.d.

For any $\beta:\alpha \in]\frac{3}{14},\infty[$ we view $T_{\beta:\alpha}$ as a tubular I-series in the categories modA, $U(\tilde{K})_{(d)}$, $\mathcal{M}$ and ${}_{\Lambda}L$. In each of these categories there is a dimension type associated with $T_{\beta:\alpha}$, which we denote by $\underline{\dim}_A T_{\beta:\alpha}$, $\underline{\dim}_U T_{\beta:\alpha}$, $\underline{\dim}_{\mathcal{M}} T_{\beta:\alpha}$ and $\underline{\dim}_\Lambda T_{\beta:\alpha}$. Of course, a similar remark applies to $\tilde{T}_\infty$. With respect to dimension types in $U(\tilde{K})_{(d)}$ and $\mathcal{M}$ we use the following notation. For

$X = (k^n, \bigoplus_{i=-3}^{8} K_i^{n_i}, \psi) \in U(\tilde{K})_{(d)}$ we write $\underline{\dim}_U X = (n; n_{-3}e_{-3}, \ldots, n_8 e_8)$.

For $M = (\bigoplus_{i=0}^{2} K_i^{n_i}, R_3^n, \varphi) \in \mathcal{M}$ we write $\underline{\dim}_{\mathcal{M}} M = (n; n_o, n_1, n_2)$.

Viewing $T_{\beta:\alpha}$ as a tubular I-series in $U(\tilde{K})_{(d)}$, we set $\mathrm{support}(T_{\beta:\alpha}) = \bigcup_{X \in T_{\beta:\alpha}} \mathrm{support}(X)$. In the sequel all intervals are understood to be intervals in $\mathbb{P}_1\mathbb{Q}$. For any two integers m and n we denote by (m,n) the greatest common divisor of m and n .

DEFINITION (The bijection $\chi :]\frac{3}{14},\infty[\rightarrowtail\!\!\!\twoheadrightarrow \mathbb{P}_1\mathbb{Q}$). We define the bijection $\chi :]\frac{3}{14},\infty[\longrightarrow \mathbb{P}_1\mathbb{Q}$ piecewise in the following way.

$$\chi\Big|_{[\frac{1}{2},\infty]} = \chi_1 : [\tfrac{1}{2},\infty] \rightarrowtail\!\!\!\twoheadrightarrow [1,\infty] \quad \text{is given by} \quad \chi_1(x) = \frac{2x}{2x-1} ,$$

$$\chi\Big|_{[\frac{1}{4},\frac{1}{2}]} = \chi_2 : [\tfrac{1}{4},\tfrac{1}{2}] \rightarrowtail\!\!\!\twoheadrightarrow [-\infty,0] \quad \text{is given by} \quad \chi_2(x) = \frac{4x-1}{2x-1} ,$$

$$\chi\Big|_{]\frac{3}{14},\frac{1}{4}]} = \chi_3 :]\tfrac{3}{14},\tfrac{1}{4}] \rightarrowtail\!\!\!\twoheadrightarrow [0,1[\quad \text{is given by} \quad \chi_3(x) = \frac{-4x+1}{10x-2} .$$

LEMMA 5.6. Let $a = (1;0,1,0)$ and $b = (2;1,0,1)$ be fixed vectors in $\mathbb{Z}^4$. Let $\beta:\alpha$ be any element in $]\frac{3}{14},\infty[$ and put $\chi(\beta:\alpha) = \mathrm{b}:\mathrm{a}$. Then $\underline{\dim}_{\mathcal{M}} T_{\beta:\alpha} = \frac{1}{(\mathrm{b},\mathrm{a})}(|\mathrm{b}|b + |\mathrm{a}|a)$. Moreover, $\underline{\dim}_{\mathcal{M}}\tilde{T}_\infty = b + a$.

PROOF. First of all, the assertion concerning $\tilde{T}_\infty$ is true, because $\underline{\dim}_{\mathcal{M}}\tilde{T}_\infty = (3;1,1,1) = b + a$.

Now let $\beta:\alpha$ be any element in $]\frac{3}{14},\infty[$. From the theory of tubular one-point extensions it is clear that

$$\underline{\dim}_{\mathrm{A}} T_{\beta:\alpha} = \begin{matrix} \alpha' & & \\ \alpha' & 2\alpha'+2\beta' & 2\beta' \\ \alpha' & & \end{matrix} ,$$

where $\alpha' = \frac{\alpha}{(2\beta,\alpha)}$ and $\beta' = \frac{\beta}{(2\beta,\alpha)}$. Consequently, $\underline{\dim}_{U} T_{\beta:\alpha} =$ $= (2\beta';\mathrm{n}_{-3}\mathrm{e}_{-3},\ldots,\mathrm{n}_8\mathrm{e}_8)$, where the numbers $\{\mathrm{n}_{-3},\ldots,\mathrm{n}_8\} \subset \mathbb{N}_0$ satisfy the condition $\underline{\dim}_{A_o}\left(\bigoplus_{i=-3}^{8} I_i^{n_i}\right) = \begin{matrix} \alpha' & \\ \alpha' & 2\alpha'+2\beta' \\ \alpha' & \end{matrix}$. This amounts to the following system of linear equations which we denote by (*).

$$\begin{aligned}
2n_{-3} + n_{-2} + 2n_{-1} + 2n_o + n_2 + 2n_3 + n_4 + n_5 &= \alpha' \\
n_{-3} + 2n_{-2} + 2n_{-1} + 2n_o + n_1 + 2n_3 + n_4 + n_6 &= \alpha' \\
n_{-3} + n_{-2} + 3n_{-1} + 2n_o + n_1 + n_2 + n_3 + n_4 + n_7 &= \alpha' \\
3n_{-3} + 3n_{-2} + 6n_{-1} + 5n_o + 2n_1 + 2n_2 + 4n_3 + 3n_4 + n_5 + n_6 + 2n_7 + n_8 &= \\
&= 2\alpha' + 2\beta'
\end{aligned}$$

Let $\beta:\alpha \in S = \{1:1, 1:2, 1:3, 1:4, 2:9\}$. Then $\underline{\dim}_U T_{\beta:\alpha}$ can easily be derived from the pattern of $\tilde{K}$.

$\beta:\alpha$	1:1	1:2	1:3	1:4	2:9
$\underline{\dim}_U T_{\beta:\alpha}$	$(2;e_5,e_6,e_7)$	$(1;e_4)$	$(2;e_1,e_2,e_3)$	$(1;e_o)$	$(4;e_{-3},e_{-2},e_{-1},2e_o)$

Passing from $\vartheta(X,Y,Z) \in U(\tilde{K})_{(d)}$ to $\sigma(X,Y,Z) \in \mathcal{M}$, we see that the assertion is true in case $\beta:\alpha \in S$.

Let $\beta:\alpha \in]1,\infty[$. Then $\mathrm{support}(T_{\beta:\alpha}) \subset \{K_5,K_6,K_7,K_8\}$, because $T_{1:1}$ and T_∞ are separating tubular series. Hence (*) has the unique solution $n_{-3} = \dots = n_4 = 0$, $n_5 = n_6 = n_7 = \alpha'$, $n_8 = 2\beta' - 2\alpha'$. Passing from $\vartheta(X,Y,Z)$ to $\sigma(X,Y,Z)$ we see that $\underline{\dim}_{\mathcal{M}} T_{\beta:\alpha} = (2\beta')b + (2\beta'-\alpha')a$. Hence the assertion is true in case $\beta:\alpha \in]1,\infty[$.

We proceed by applying the same argument successively.

Let $\beta:\alpha \in]\frac{1}{2},1[$. Then $\mathrm{support}(T_{\beta:\alpha}) \subset \{K_4,K_5,K_6,K_7\}$ and therefore (*) has the unique solution $n_{-3} = \dots = n_3 = n_8 = 0$, $n_4 = -2\beta' + 2\alpha'$, $n_5 = n_6 = n_7 = 2\beta' - \alpha'$. Hence $\underline{\dim}_{\mathcal{M}} T_{\beta:\alpha} = (2\beta')b + (2\beta'-\alpha')a$, which proves the assertion in case $\beta:\alpha \in]\frac{1}{2},1[$.

Let $\beta:\alpha \in]\frac{1}{3},\frac{1}{2}[$. Then $\mathrm{support}(T_{\beta:\alpha}) \subset \{K_1,K_2,K_3,K_4\}$ and therefore (*) has the unique solution $n_{-3} = \ldots = n_o = n_5 = \ldots = n_8 = 0$, $n_1 = n_2 = n_3 = -2\beta' + \alpha'$, $n_4 = 6\beta' - 2\alpha'$. Hence $\underline{\dim}_{\mathcal{M}} T_{\beta:\alpha} = (4\beta'-\alpha')b + (-2\beta'+\alpha')a$, which proves the assertion in case $\beta:\alpha \in]\frac{1}{3},\frac{1}{2}[$.

Let $\beta:\alpha \in]\frac{1}{4},\frac{1}{3}[$. Then $\mathrm{support}(T_{\beta:\alpha}) \subset \{K_o,K_1,K_2,K_3\}$ and therefore (*) has the unique solution $n_{-3} = \ldots = n_{-1} = n_4 = \ldots = n_8 = 0$, $n_o = -6\beta' + 2\alpha'$, $n_1 = n_2 = n_3 = 4\beta' - \alpha'$. Hence $\underline{\dim}_{\mathcal{M}} T_{\beta:\alpha} = (4\beta'-\alpha')b + (-2\beta'+\alpha')a$, which proves the assertion in case $\beta:\alpha \in]\frac{1}{4},\frac{1}{3}[$.

Let $\beta:\alpha \in]\frac{3}{14},\frac{1}{4}[$. Then $\mathrm{support}(T_{\beta:\alpha}) \subset \{K_{-3},K_{-2},K_{-1},K_o\}$ and (*) has the unique solution $n_{-3} = n_{-2} = n_{-1} = -4\beta' + \alpha'$, $n_o = 10\beta' - 2\alpha'$, $n_1 = \ldots = n_8 = 0$. However, in the present situation this information is not sufficient in order to derive $\underline{\dim}_{\mathcal{M}} T_{\beta:\alpha}$. Namely for all $X \in T_{\beta:\alpha}$ and all $i = -3,-2,-1$ we have the partition $n_i = m_4 + m_{i+8}$, where the numbers $m_4,\ldots,m_7$ arise from the decomposition $A_{-1}(X) = \bigoplus_{i=4}^{8} I_i^{m_i}$. We claim that $m_4 = 0$ if $\beta:\alpha > \frac{2}{9}$ and that $m_8 = 0$ if $\beta:\alpha < \frac{2}{9}$.

Suppose that $\beta:\alpha > \frac{2}{9}$ and that $m_4 > 0$ for some $X \in T_{\beta:\alpha}$. Then there exists a factorobject $X/X_1 = A_{-1}(X/X_1) = I_4$ of X . Hence $U(\tilde{K})(X,X/X_1) \neq 0$. On the other hand, $\gamma(X/X_1) = \frac{2}{9}$ and therefore $U(\tilde{K})(X,X/X_1) = 0$, which is a contraction.

Suppose that $\beta:\alpha < \frac{2}{9}$ and let $X = (U,K,\psi)$ be any object in $T_{\beta:\alpha}$. Consider the subobject $X_2 = (U_2,K_o^{n_o},\psi_2)$ of X , where $U_2 = \ker(\alpha_{-1}\psi) \cong k^{m_8}$, $a_{-1} : K \twoheadrightarrow \bigoplus_{i=-3}^{-1} K_i^{n_i}$ being the projection,

$\psi_2 = \psi|_{U_2}$. Then for all indecomposable direct summands N of X_2 we have $U(\tilde{K})(N,X) \neq 0$. On the other hand, $\gamma(N) \geq \frac{2}{9}$ and therefore $U(\tilde{K})(N,X) = 0$, for all indecomposable direct summands N of X_2 which are different from $(0,K_o,o)$. Hence $X_2 = (0,K_o^{n_o},o)$ and $m_8 = 0$.

This information on m_4 and m_8 , together with the solution of (*), gives the equation $\underline{\dim}_{\mathcal{M}} T_{\beta:\alpha} = (-4\beta'+\alpha')b = (10\beta'-2\alpha')a$, for all $\beta:\alpha \in]\frac{3}{14},\frac{1}{4}[$. This proves the assertion in case $\beta:\alpha \in]\frac{3}{14},\frac{2}{9}[\cup]\frac{2}{9},\frac{1}{4}[$. q.e.d.

We conclude that the proof of Theorem II is now complete. Indeed, from Corollary 5.5 we obtain that $\mathcal{A}_s(\Lambda) \cong \mathcal{A}_s(\Lambda)^{op} \cong \dot{\bigcup}_{\frac{3}{14}<\beta:\alpha\leq\infty} T_{\beta:\alpha}$, where we set $T_\infty := \tilde{T}_\infty$. Replacing the indices $\beta:\alpha \in]\frac{3}{14},\infty]$ by indices $\chi(\beta:\alpha) = b:a \in \mathbb{P}_1\mathbb{Q}$, we obtain that $\mathcal{A}_s(\Lambda) \cong \dot{\bigcup}_{b:a\in\mathbb{P}_1\mathbb{Q}} T_{b:a}$, proving (i). For all $b:a \in \mathbb{P}_1\mathbb{Q}$, the tubular type of $T_{b:a}$ is $\tilde{\mathbb{CD}}_3$, proving (ii). From Lemma 5.6 we deduce that for all $b:a \in \mathbb{P}_1\mathbb{Q}$, $\underline{\dim}_{\mathcal{M}} T_{b:a} = \frac{1}{(b,a)}(|b|\ \underline{\dim}_{\mathcal{M}} T_{1:0} + |a|\ \underline{\dim}_{\mathcal{M}} T_{0:1})$ and hence $\underline{\dim}_{\Lambda} T_{b:a} = \frac{1}{(b,a)}(|b|\ \underline{\dim}_{\Lambda} T_{1:0} + |a|\ \underline{\dim}_{\Lambda} T_{0:1}) = \frac{1}{(b,a)}(6|b| + 3|a|)$, proving (iii). Finally it is easy to see that $\mathrm{rad}\Lambda$ is an indecomposable Λ-lattice, and that its isomorphism class belongs to the mouth of the stable tube $T_{0:1}(\infty)$ of $\mathcal{A}_s(\Lambda)$, which proves (iv).

6. APPENDIX

In fact we have proved Theorem I and Theorem II in a more explicit sense than becomes apparent from their statements. Since now the full machinery is at our disposal, it seems appropriate to reformulate the Theorems, taking account of the constructive aspects of the proof. Section 6.1 may be considered as a reformulation of Theorem I and section 6.2 as a reformulation of Theorem II. Section 6.3 contains a Leitfaden for the entire proof.

6.1. COMPLETE LIST OF ALL INDECOMPOSABLE Λ-LATTICES.

Tracing backwards the four reductions which constitute the proof of Theorem II we give a complete "list" of the isomorphism classes of indecomposable Λ-lattices and we demonstrate the effectivity of the bijection asserted in Theorem I.

Let M be an indecomposable A-module in $\dot{\bigcup}_{\frac{3}{14}<\beta:\alpha<\infty} T_{\beta:\alpha} \mathbin{\dot{\cup}} \check{T}_{\infty}$, where $\check{T}_{\infty} = \dot{\bigcup}_{\lambda\in\check{I}} T_{\infty}(\lambda)$. (For definition of A and description of $A(\mathrm{A})$ see discussion ensuing Lemma 5.1.) Let $X = X(\mathrm{M})$ be the object in $U(\tilde{K})$ which corresponds to M under the functor $\Psi_1\Psi_0 : Q_0 \to U(\tilde{K})$. (For a description of this functor see discussion preceding Lemma 5.2.) Then $X \in \operatorname{ind}U(\tilde{K})_{(\mathrm{d})}$, by Proposition 5.4. There exists an angular matrix $(\mathrm{X,Y,Z}) \in \mathfrak{M}$ such that $X \cong \vartheta(\mathrm{X,Y,Z})$, by Lemma 5.3. (The determination of $(X,Y,Z) \in \mathfrak{M}$ for a given object $X \in \mathrm{ind}\,U(\tilde{K})_{(d)}$ amounts to an easy matrix problem.) From $\vartheta(X,Y,Z) \in \hat{D}$ we pass to $\zeta(X,Y,Z) \in \hat{C}$, to $\Psi(\zeta(X,Y,Z)) \in [\hat{C}]$ and finally to $\sigma(X,Y,Z) = \phi^{-1}\Psi(\zeta(X,Y,Z)) \in \hat{S}$. (For definition of the mappings ϑ, ζ and σ see section 0.4.) The latter matrix is of the form $\sigma(X,Y,Z) = (\bar{A}_o \| \pi^2\bar{A}_1 | \bar{A}_2 \| \pi\bar{A}_3 | \pi\bar{A}_4)$, where $\bar{A}_i \in R_3^{m\times n_i}$, $i = 0,\dots,4$, subject to $n_1 = n_2$ and $n_3 = n_4$. For all $i = 0,\dots,4$ let $A_i \in R_4^{m\times n_i}$, subject to $\rho_3(A_i) = \bar{A}_i$, where ρ_3 denotes reduction modulo π^3. Then $(\pi A_o \| \pi^3 A_1 | \pi A_2 \| \pi^2 A_3 | \pi^2 A_4) = \phi_1^{-1}(\sigma(X,Y,Z)) \in \check{F}_o$. For all $i = 0,\dots,4$ let $\tilde{A}_i \in R^{m\times n_i}$, subject to $\rho_4(\tilde{A}_i) = A_i$, where ρ_4 denotes reduction modulo π^4. Set

$$\tilde{\alpha}_s^{n_s} = \left(\begin{array}{c|c} E_{n_s} & d\pi^{4-s}E_{n_s} \\ \hline \pi^s E_{n_s} & -2E_{n_s} \end{array}\right)$$

for all $s = 0,1,2$, and define $A \in R^{n\times n}$, with $n = m + 2(n_o+n_1+n_2)$, by

$$A = \left(\begin{array}{c|cc|cc|cc} E_m & 0 & \pi\tilde{A}_o & \tilde{A}_1 & d\tilde{A}_2 & \tilde{A}_3 & d\tilde{A}_4 \\ \hline & \multicolumn{2}{c|}{\tilde{\alpha}_o^{n_o}} & & & & \\ \cline{2-5} 0 & & & \multicolumn{2}{c|}{\tilde{\alpha}_1^{n_1}} & & \\ \cline{4-7} & & & & & \multicolumn{2}{c}{\tilde{\alpha}_2^{n_2}} \end{array}\right)$$

The free R-module $L = R^n$, together with the endomorphism given by A, is a Λ-lattice since $A^3 = E_n$, and in fact $L = \phi_o^{-1}\phi_1^{-1}(\sigma(X,Y,Z)$. Due to the proof of Theorem II, L is indecomposable and its isomorphism class is uniquely determined by the isomorphism class of M. We write $L = L(M)$ and $\mathcal{L}_1 = \{L(M)\}$, as M ranges through a complete set of representatives for the isomorphism classes in $\dot{\bigcup}_{\frac{3}{14}<\beta:\alpha<\infty} \mathcal{T}_{\beta:\alpha} \,\dot{\cup}\, \check{\mathcal{T}}_\infty$.

Also, let T be an indecomposable object in $\mathcal{T}(E)$. (For a classification of $\mathcal{T}(E)$ see Proposition 3.4 together with its preceding list.) It is of the form $T = (\bar{A}_o \| \pi^2\bar{A}_1 | \bar{A}_2 \| \pi\bar{A}_3 | \pi\bar{A}_4)$, where $\bar{A}_i \in R_3^{m\times n_i}$, $i = 0,\ldots,4$, subject to $n_1 = n_2$ and $n_3 = n_4$. The Λ-lattice $L = \phi_o^{-1}\phi_1^{-1}(T)$, constructed as above, is indecomposable and its isomorphism class is uniquely determined by the isomorphism class of T. We write $L = L(T)$ and $\mathcal{L}_2 = \{L(T)\}$ as T ranges through a complete set of representatives for the isomorphism classes in $\operatorname{ind}\mathcal{T}(E)$.

Now we may reformulate one part of Theorem I as follows.

<u>THEOREM I'</u>. All Λ-lattices in $\mathcal{L}_1 \cup \mathcal{L}_2$ are indecomposable and pairwise nonisomorphic. Every indecomposable nonprojective Λ-lattice is isomorphic to a lattice in $\mathcal{L}_1 \cup \mathcal{L}_2$.

Moreover, we give an explicit description of the bijection asserted in Theorem I. Set $[\mathcal{L}_1 \cup \mathcal{L}_2] = \{[L] \mid L \in \mathcal{L}_1 \cup \mathcal{L}_2\}$, where $[L]$ denotes the isomorphism class of L. Let $(\ ) : \bar{I} \twoheadrightarrow I$ be the canonical surjection which is given by $(\lambda) = \lambda$ for all $\lambda \in I$, and

$(\bar{\lambda}) = \lambda$ for all $\lambda \in E$. Let J be the distinguished subset $J = \{(1{:}1,\lambda,n)\ ,\ (1{:}1),\bar{\lambda},n) \mid \lambda \in E,\ n \in \mathbb{N}\}$ of $\mathbb{P}_1\mathbb{Q} \times \bar{I} \times \mathbb{N}$. We define a mapping $\varphi\colon \mathbb{P}_1\mathbb{Q} \times \bar{I} \times \mathbb{N} \longrightarrow [L_1 \cup L_2]$ in the following way. If $(\beta{:}\alpha,\lambda,n) \in (\mathbb{P}_1\mathbb{Q} \times \bar{I} \times \mathbb{N})\setminus J$ then choose an indecomposable A-module $M(\beta{:}\alpha,\lambda,n)$ representing the n-th point on the ray λ of the component $T_{\chi^{-1}(\beta:\alpha)}((\lambda))$ of $A(A)$, and define $\varphi(\beta{:}\alpha,\lambda,n) = [L(M(\beta{:}\alpha,\lambda,n))]$. If $(1{:}1,\varepsilon,n) \in J$ then let $T(\varepsilon)_n$ be an indecomposable object in $T(E)$ representing the n-th point on the ray ε of the component $A_{\mathcal{M}}(T((\varepsilon)))$ of $A(\mathcal{M})$, and define $\varphi(1{:}1,\varepsilon,n) = [L(T(\varepsilon)_n)]$. (For definition of the bijection $\chi :]\frac{3}{14},\infty] \rightarrowtail\!\!\!\twoheadrightarrow \mathbb{P}_1\mathbb{Q}$ see section 5. For description of $A_{\mathcal{M}}(T(\varepsilon))$ see Proposition 3.4.)

Then we may reformulate the other part of Theorem I as follows.

<u>THEOREM I"</u>. The mapping $\varphi\colon \mathbb{P}_1\mathbb{Q} \times \bar{I} \times \mathbb{N} \longrightarrow [L_1 \cup L_2]$ is a bijection.

6.2. THE AUSLANDER-REITEN QUIVER OF $\mathcal{M}$.

In view of Corollary 1.3, Corollary 5.5 and Lemma 5.6 we also have determined the Auslander-Reiten quiver of $\mathcal{M}$. It is worth while to put this on record, because the knowledge of $A(\mathcal{M})$ will refine some of the information contained in Theorem II. There are two reasons for this feature. Firstly, $\underline{\dim}_{\mathcal{M}}$ has values in $\mathbb{Z}^4$ whereas $\underline{\dim}_{\Lambda}$ has values in $\mathbb{Z}$, and secondly, the quiver isomorphism $A(\mathcal{M}) \cong A_s(\Lambda)$ from Corollary 1.3 is not an isomorphism of translation quivers.

Indeed, there are eight points in $A_s(\Lambda)$ which become unstable in

$A(\mathcal{M})$, four of them becoming projective points and four of them becoming injective points (see statement (2) at the beginning of section 3):

$$P_\infty = (1) \ , \ \underline{\dim}_{\mathcal{M}} P_\infty = (1;1,0,0) \ ; \ I_\infty = (0|1) \ , \ \underline{\dim}_{\mathcal{M}} I_\infty = (1;0,1,0);$$

$$P_o = \mathrm{I} \ , \ \underline{\dim}_{\mathcal{M}} P_o = (1;0,0,0) \ ; \ I_o = \vdash\!\!\dashv \ , \ \underline{\dim}_{\mathcal{M}} I_o = (0;1,0,0);$$

$$P_\delta = \begin{pmatrix} \pi & 0 \\ 0 & \pi \end{pmatrix} , \ \underline{\dim}_{\mathcal{M}} P_\delta = (2;0,0,1) \ ; \ I_\delta = \vdash\!\!-\!\!+\!\!-\!\!\dashv \ , \ \underline{\dim}_{\mathcal{M}} I_\delta = (0;0,0,1);$$

$$P^\infty = \begin{pmatrix} \pi^2 & 0 \\ 0 & 1 \end{pmatrix} , \ \underline{\dim}_{\mathcal{M}} P^\infty = (2;0,1,0) \ ; \ I^\infty = \vdash\!\!-\!\!+\!\!-\!\!\dashv \ , \ \underline{\dim}_{\mathcal{M}} I^\infty = (0;0,1,0).$$

Writing $p_\infty = [P_\infty]$, $i_\infty = [I_\infty]$, etc., we have the following complete description of the Auslander-Reiten quiver of $\mathcal{M}$.

<u>THEOREM II'</u>. (i) The stable Auslander-Reiten quiver of $\mathcal{M}$ is given by a $\mathbb{P}_1\mathbb{Q}$-family of tubular I-series: $A_s(\mathcal{M}) \cong \dot{\bigcup}_{\beta:\alpha\in\mathbb{P}_1\mathbb{Q}} T_{\beta:\alpha}$.

(ii) The tubular type of $T_{\beta:\alpha}$ does not depend on $\beta:\alpha \in \mathbb{P}_1\mathbb{Q}$. It only depends on δ being reducible in $k[X]$ or not. If δ is reducible in $k[X]$ then the tubular type is $\tilde{\mathbb{D}}_4$. If δ is irreducible in $k[X]$ then the tubular type is $\widetilde{\mathbb{CD}}_3$.

(iii) For all $\beta:\alpha \in \mathbb{P}_1\mathbb{Q}$, the dimension type of $T_{\beta:\alpha}$ is given by

$$\underline{\dim}_{\mathcal{M}} T_{\beta:\alpha} = |\beta| \underline{\dim}_{\mathcal{M}} T_{1:0} + |\alpha| \underline{\dim}_{\mathcal{M}} T_{0:1} = (2|\beta|+|\alpha|; |\beta|, |\alpha|, |\beta|) ,$$

where (β,α) is a pair of relatively prime integers representing $\beta:\alpha$.

(iv) The position of the unstable points p_∞, p_o, p_δ, p^∞ and i_∞, i_o, i_δ, i^∞ of $A(\mathcal{M})$ are at the mouths of the exceptional tubes $T_{0:1}(\infty)$, $T_{0:1}(0)$, $T_{0:1}(\delta)$ and $T_{1:0}(\infty)$:

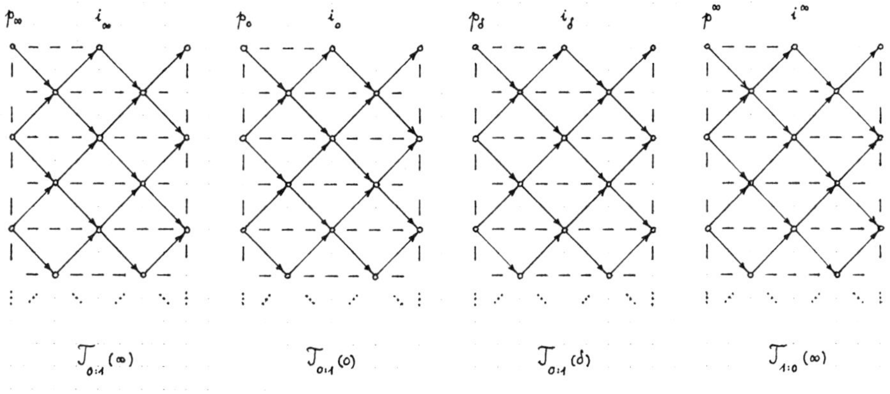

(The vertical interrupted lines indicate the Auslander-Reiten translation.)

Motivated by assertions (i) and (iii), and by linear independence of the vectors $\underline{\dim}_{\mathcal{M}} T_{0:1} = (1;0,1,0)$ and $\underline{\dim}_{\mathcal{M}} T_{1:0} = (2;1,0,1)$, we try to visualize the Auslander-Reiten quiver of $\mathcal{M}$ by the picture below. Each of the tubular I-series $T_{\beta:\alpha}$ is represented by a half line. The family $\{T_{\beta:\alpha}\}_{\beta:\alpha>0}$ constitutes the front side and the family $\{T_{\beta:\alpha}\}_{\beta:\alpha<0}$ constitutes the back side of the cone. The six tubular I-series $T_{1:1}$, $T_{2:1}$, $T_{1:0}$, $T_{-1:1}$, $T_{0:1}$, $T_{1:2}$ which "span" the Auslander-Reiten quiver of $\mathcal{M}$ (in the sense of the proof of Lemma 5.6) are shown in bold print and are marked by their indices.

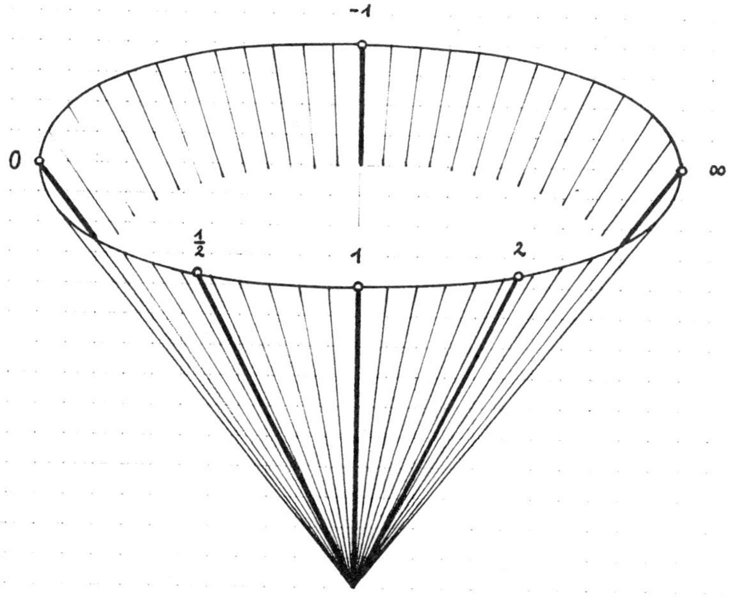
-1
0
∞
1/2
1
2

6.3. LEITFADEN.

We write $F_o = F(K_o)$ and $F_1 = F(K_1)$. There occur four types of functors for which we use the following symbols. Given any Krull-Schmidt categories K and K' ,

$K \cong K'$ denotes an equivalence,

$K \stackrel{\scriptscriptstyle\frown}{=} K'$ denotes a duality,

$K \simeq K'$ denotes a representation equivalence which induces an isomorphism $A(K) \cong A(K')$,

$K \hookrightarrow K'$ denotes an embedding which induces an isomorphism $A(K) \cong A_{K'}(K)$.

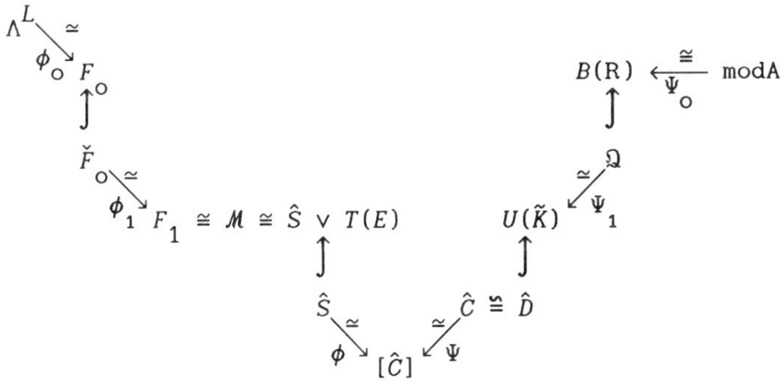

REFERENCES

[Au75] M. Auslander: Existence theorems for almost split sequences. Ring theory II, Proc. Second Oklahoma Conference, Marcel Dekker 1975, 1-44.

[Au76] M. Auslander: Functors and morphisms determined by objects. Representation theory of algebras. Proc. Philadelphia Conference, Marcel Dekker 1976, 1-244.

[Au84] M. Auslander: Isolated singularities and existence of almost split sequences. Notes by L. Unger. Proceedings of the fourth international conference on representations of algebras, Carleton University, Ottawa 84. Springer Lecture Notes 1178, 194-242.

[Di83a] E. Dieterich: Construction of Auslander-Reiten quivers for a class of group rings. Math. Z. 184 (1983), 43-60.

[Di83b] E. Dieterich: Group rings of wild representation type. Math. Ann. 266 (1983), 1-22.

[Di85] E. Dieterich: Lattices over group rings of cyclic p-groups and generalized factorspace categories. Journal of the London Math. Soc. (2)31 (1985), 407-424.

[Dl/Ri76] V. Dlab/C.M. Ringel: Indecomposable representations of graphs and algebras. Memoirs American Math. Soc. 173 (1976).

[Dr/Ro67] J.A. Drozd/A. Roiter: Commutative rings with a finite number of indecomposable integral representations. Izv. Akad. Nauk SSSR, Ser. Mat. 31, 783-798 (1967) (Russian). English translation: Math. USSR Izv. 1, 757-772 (1967).

[Ja67] H. Jacobinski: Sur les ordres commutatifs avec une nombre fini de réseaux indecomposables. Acta Math. 118, 1-31 (1967).

[Ja72] A.V. Jacovlev: Classification of the 2-adic representations of the cyclic group of order eight. Zap. Naučn. Sem. Leningrad. Otdel. Mat. Inst. Steklov. (LOMI), 28 (1972), 93-129; Journal Soviet Math. 3 (1975), 654-680.

[Ko75] N.M. Kopelevič: Representations of the cyclic group of order four over the ring of Gaussian integers. Manuscript 514-775, deposited at VINITI by the editors of Vestnik Leningrad. Univ. Mat. Meh. Astronom. (1975) (Russian).

[Na67] L.A. Nazarova: Representations of a tetrad. Izv. Akad. Nauk SSR Ser. Mat. 31 (1967), 1361-1378; Mat. USSR Izv. 1, 1305-1321 (1967).

[Ri79] C.M. Ringel: Tame algebras. Proceedings of the Ottawa conference in representation theory (Ottawa 1979), 137-287. Springer Lecture Notes 831 (1980).

[Ri84] C.M. Ringel: Tame algebras and integral quadratic forms. Springer Lecture Notes 1099 (1984).

[RoS76] K.W. Roggenkamp/J.W. Schmidt: Almost split sequences for integral group rings and orders. Comm. Algebra 4, 893-917 (1976).

[Wi80] A. Wiedemann: Orders with loops in their Auslander-Reiten graph. Comm. Algebra 9, 641-656 (1981).

Ernst Dieterich
Universität Bielefeld
Fakultät für Mathematik
Universitätsstrasse
D-4800 Bielefeld 1
West Germany.

Current address:
Ernst Dieterich
Mathematisches Institut
Universität Zürich
Rämistrasse 74
CH-8001 Zürich
Switzerland.

MEMOIRS of the American Mathematical Society

SUBMISSION. This journal is designed particularly for long research papers (and groups of cognate papers) in pure and applied mathematics. The papers, in general, are longer than those in the TRANSACTIONS of the American Mathematical Society, with which it shares an editorial committee. Mathematical papers intended for publication in the Memoirs should be addressed to one of the editors:

Ordinary differential equations, partial differential equations and applied mathematics to ROGER D. NUSSBAUM, Department of Mathematics, Rutgers University, New Brunswick, NJ 08903

Harmonic analysis, representation theory and Lie theory to AVNER D. ASH, Department of Mathematics, The Ohio State University, 231 West 18th Avenue, Columbus, OH 43210

Abstract analysis to MASAMICHI TAKESAKI, Department of Mathematics, University of California, Los Angeles, CA 90024

Real and harmonic analysis to DAVID JERISON, Department of Mathematics, M.I.T., Rm 2-180, Cambridge, MA 02139

Algebra and algebraic geometry to JUDITH D. SALLY, Department of Mathematics, Northwestern University, Evanston, IL 60208

Geometric topology and general topology to JAMES W. CANNON, Department of Mathematics, Brigham Young University, Provo, UT 84602

Algebraic topology and differential topology to RALPH COHEN, Department of Mathematics, Stanford University, Stanford, CA 94305

Global analysis and differential geometry to JERRY L. KAZDAN, Department of Mathematics, University of Pennsylvania, E1, Philadelphia, PA 19104-6395

Probability and statistics to RICHARD DURRETT, Department of Mathematics, Cornell University, Ithaca, NY 14853-7901

Combinatorics and number theory to CARL POMERANCE, Department of Mathematics, University of Georgia, Athens, GA 30602

Logic, set theory, general topology and universal algebra to JAMES E. BAUMGARTNER, Department of Mathematics, Dartmouth College, Hanover, NH 03755

Algebraic number theory, analytic number theory and modular forms to AUDREY TERRAS, Department of Mathematics, University of California at San Diego, La Jolla, CA 92093

Complex analysis and nonlinear partial differential equations to SUN-YUNG A. CHANG, Department of Mathematics, University of California at Los Angeles, Los Angeles, CA 90024

All other communications to the editors should be addressed to the Managing Editor, DAVID J. SALTMAN, Department of Mathematics, University of Texas at Austin, Austin, TX 78713.

General instructions to authors for

PREPARING REPRODUCTION COPY FOR MEMOIRS

For more detailed instructions send for AMS booklet, "A Guide for Authors of Memoirs." Write to Editorial Offices, American Mathematical Society, P.O. Box 6248, Providence, R.I. 02940.

MEMOIRS are printed by photo-offset from camera copy fully prepared by the author. This means that the finished book will look exactly like the copy submitted. Thus the author will want to use a good quality typewriter with a new, medium-inked black ribbon, and submit clean copy on the appropriate model paper.

Model Paper, provided at no cost by the AMS, is paper marked with blue lines that confine the copy to the appropriate size.

Special Characters may be filled in carefully freehand, using dense black ink, or **INSTANT** ("rub-on") **LETTERING** may be used. These may be available at a local art supply store.

Diagrams may be drawn in black ink either directly on the model sheet, or on a separate sheet and pasted with rubber cement into spaces left for them in the text. Ballpoint pen is not acceptable.

Page Headings (Running Heads) should be centered, in CAPITAL LETTERS (preferably), at the top of the page — just above the blue line and touching it.

LEFT-hand, EVEN-numbered pages should be headed with the AUTHOR'S NAME;

RIGHT-hand, ODD-numbered pages should be headed with the TITLE of the paper (in shortened form if necessary).

Exceptions: PAGE 1 and any other page that carries a display title require NO RUNNING HEADS.

Page Numbers should be at the top of the page, on the same line with the running heads.

LEFT-hand, EVEN numbers — flush with left margin;

RIGHT-hand, ODD numbers — flush with right margin.

Exceptions: PAGE 1 and any other page that carries a display title should have page number, centered below the text, on blue line provided.

FRONT MATTER PAGES should be numbered with Roman numerals (lower case), positioned below text in same manner as described above.

MEMOIRS FORMAT

It is suggested that the material be arranged in pages as indicated below.
Note: Starred items (*) are requirements of publication.

Front Matter (first pages in book, preceding main body of text).

Page i — *Title, *Author's name.

Page iii — Table of contents.

Page iv — *Abstract (at least 1 sentence and at most 300 words).

Key words and phrases, if desired. (A list which covers the content of the paper adequately enough to be useful for an information retrieval system.)

**1991 Mathematics Subject Classification*. This classification represents the primary and secondary subjects of the paper, and the scheme can be found in Annual Subject Indexes of MATHEMATICAL REVIEWS beginnning in 1990.

Page 1 — Preface, introduction, or any other matter not belonging in body of text.

Footnotes: *Received by the editor date.
Support information — grants, credits, etc.

First Page Following Introduction - Chapter Title (dropped 1 inch from top line, and centered). Beginning of Text.

Last Page (at bottom) - Author's affiliation.